Computational Modeling of Pulverized Coal Fired Boilers

Vivek V. Ranade
Devkumar F. Gupta

CRC Press
Taylor & Francis Group
Boca Raton London New York

CRC Press is an imprint of the
Taylor & Francis Group, an **informa** business

Computational Modeling of Pulverized Coal Fired Boilers

CRC Press
Taylor & Francis Group
6000 Broken Sound Parkway NW, Suite 300
Boca Raton, FL 33487-2742

First issued in paperback 2017

© 2015 by Taylor & Francis Group, LLC
CRC Press is an imprint of Taylor & Francis Group, an Informa business

No claim to original U.S. Government works

ISBN-13: 978-1-4822-1528-1 (hbk)
ISBN-13: 978-1-138-74842-2 (pbk)

Visit the Taylor & Francis Web site at
http://www.taylorandfrancis.com

and the CRC Press Web site at
http://www.crcpress.com

Contents

Preface

Pulverized coal fired boilers have been and will be the mainstay of coal-based power generation worldwide. Such pulverized coal boilers are complex chemical reactors comprising many processes such as gas-solid flows, combustion, heat transfer (conduction, convection, and radiation), and phase change simultaneously. Conventional pulverized coal fired boiler designs have been continuously improved upon and even today new combustion technologies are being developed. Improving the efficiency of pulverized coal (PC) fired boilers has been the focus of considerable efforts by the utility industry since it leads to several benefits (reduced emissions and consumption of coal per MWe). Typically, a one percent improvement in overall efficiency can result in a nearly three percent reduction in CO_2 emission.

Efficiency of these PC boilers depends on several issues such as flow and mixing, inter-phase and intra-particle heat and mass transfer, and homogeneous and heterogeneous reactions. Some specific issues are critical in determining overall efficiency of such boilers. These are coal composition (proximate and ultimate analysis), characteristics of pulverized coal particles (shape and size distributions), reactivity of coal (devolatalization and combustion kinetics of coal), boiler design and configuration (size and shape of combustion chamber, burner design, number and locations of burners), excess air used, mixing of air and coal, flow mal-distribution, generation of fly and bottom ash particles, radiative heat transfer, heat recovery, and so on. Any efforts in improving the effective operating efficiency of these boilers therefore rest on fundamental understanding and control of the underlying flow, mixing, heat transfer, and reactions.

Computational modeling offers an excellent way to develop such fundamental understanding and to develop an ability to optimize boiler performance. However, through our interactions with industry, we realized that there is insufficient help available to practicing engineers for harnessing state-of-the-art computational modeling tools for complex, industrial boiler-like applications. Many boiler engineers either consider the complexities of industrial PC boilers impossible to simulate or expect miracles from off-the-shelf, commercial modeling tools. These two diverse views arise because of inadequate understanding about the role, state of the art, and possible limitations of computational modeling.

This book is aimed at filling this gap and providing a detailed account of the methodology of computational modeling of pulverized coal boilers. The book attempts to develop and discuss an appropriate approach to model complex processes occurring in PC boilers in a tractable way. The scope is restricted to the combustion side of the boiler. The rest of the components of PC boilers are outside the scope. Even for the combustion side, the scope is restricted to the burner/excess air entry points to the exit of flue gases (and fly ash) from the heat recovery section. Milling and conveying of coal particles as well as further processing of flue gas beyond heat recovery sections (electro-static precipitators and so on) are beyond the scope of the present work. We have written this book with an intention to describe the individual aspects of combustion and heat recovery sections of PC boilers in a coherent fashion that may be useful to further improve design methodologies and optimize boiler performance. The intended users of this book are practicing engineers working in utility industries and in boiler design companies, as industrial consultants, in R & D laboratories, as well as engineering scientists/research students. Some prior background of reaction engineering and numerical techniques is assumed.

The information in the book is organized in mainly six chapters: the first chapter provides a general introduction. The second chapter discusses the overall approach and methodology. Kinetics of coal pyrolysis (devolatilization) and combustion and methods of its evaluation are discussed in Chapter 3. Computational flow modeling is discussed in Chapter 4. This chapter covers modeling aspects from the formulation of model equations to simulation methodology. Typical results obtained with computational flow models are also discussed in this chapter. Computational flow models provide a framework for developing a deeper understanding of the underlying processes in PC boilers. These models also provide a quantitative relationship between boiler hardware and operating protocols with boiler performance and efficiency. Phenomenological models or reactor network models are discussed in Chapter 5. These models require significantly less computational resources than computational flow models and therefore can be used for boiler optimization. A brief discussion on practical applications of computational modeling is included in Chapter 6. An attempt is made here to provide specific comments to connect modeling with real-life applications. Key points are summarized in the last chapter (Chapter 7) along with some comments about the path forward.

Though many of the examples are for the older generation 210 MWe boiler, we have made an attempt to evolve general guidelines that will be useful for solving practical problems related to current and future generations of PC boilers. The material presented here can be extended to model larger boilers based on conventional, super-critical, or ultra-super-critical technologies as well as based on oxy-fuel technologies. The material included in this book may be used in several ways: as a basic resource of methodologies or in making decisions about applications of computational modeling in practice. The content could be useful as a study material for an in-house course on pulverized coal boilers, for example, design and optimization or a companion book while solving practical boiler-related problems. We hope that this book will encourage chemical engineers to exploit the potential of computational models for better engineering of pulverized coal boilers.

We are grateful to many of our associates and collaborators with whom we worked on different industrial projects. Many of VVR's students have contributed to this book in different ways. Particularly, Ankit Jain, Akshay Singan, Ajinkya Pandit, Deepankar Sharma, and Dhananjay Mote read the early drafts of manuscript and provided useful suggestions from a student's perspective. Vinayak Ranade, Nanda Ranade, and Maya Gupta also read the early drafts and suggested ways to improve overall readability. The manuscript was improved wherever their suggestions were incorporated. Any remaining errors or shortcomings are, needless to say, the responsibility of the authors. We also wish to thank the editorial team at CRC Press, particularly Dr. Gagandeep Singh, for their patience and help during the process of writing this book.

Besides the professional associates, VVR would like to acknowledge his wife Nanda and daughter Vishakha, and DFG would like to acknowledge his wife Maya and daughters Suhani and Anjani for their understanding and enthusiastic support all along.

Vivek V. Ranade

Devkumar F. Gupta
Pune, India

About the Authors

Vivek V. Ranade is a deputy director of CSIR-National Chemical Laboratory, Pune, India (www.ncl-india.org) and leads chemical engineering and process development research. His research interests include process intensification, reactor engineering, and multiphase flows. He has successfully developed solutions for enhancing the performance of a wide range of industrial processes and has facilitated their implementation in practice. He has also developed various devices (micro-reactors, filters, vortex diodes) and products, and has published more than 120 papers and 3 books. Dr. Ranade is a recipient of several awards, including the Shanti Swarup Bhatnagar award and the DST Swarna Jayanti Fellowship. He is a fellow of the Indian National Academy of Engineering and the Indian Academy of Sciences. He is also an entrepreneur and co-founded a company, Tridiagonal Solutions (www.tridiagonal.com). He is currently leading a large program on process intensification, titled Indus MAGIC (www.indusmagic.org), which is aimed at developing MAGIC (modular, agile, intensified, and continuous) processes and plants. He holds B.Chem.Eng. (1984) and PhD (1988) degrees from the University of Mumbai (UDCT), India.

Devkumar F. Gupta is currently a principal scientist at Thermax Limited, Pune, India. His research interest lies in converting solid fuels into high-value-added products using fluidized bed reactors. He has been actively involved and had an instrumental role in the design, computational modeling, and establishment of high-pressure gasification demonstration plants in India. He has worked jointly with utility industries to develop computational fluid dynamics (CFD) and reduced order numerical models for pulverized coal fired boilers. He has presented several research papers at international and national conferences on clean coal technology. He holds a PhD (2011) in chemical engineering from the University of Pune.

1

Introduction

Pulverized coal (PC) fired boilers have been the mainstay of coal-based power generation worldwide for almost 100 years. This is not surprising as coal is the most abundant and widely distributed fossil fuel, with global reserves of about 1,000 billion tons (IEA, 2010a). Coal fuels more than 40% of the world's electricity. The percentage of coal-based electricity is much higher in some countries, such as South Africa (93%), Poland (92%), China (79%), Australia (78%), Kazakhstan (75%), and India (69%). The growing needs of developing countries are likely to ensure that coal remains a major source of electricity in the foreseeable future despite climate change policies (IEA, 2010 a, b).

Electricity generation from coal appears to be a rather simple process. In most coal-fired power plants, coal is first milled to a fine powder (pulverized coal). Fine particles of coal have more surface area and therefore burn more effectively. In these systems, called pulverized coal (PC) combustion systems, the powdered coal is fed to the combustion chamber of a boiler. Coal particles burn in this chamber and generate a fireball (high-temperature zone) and hot gases. The energy released in radiative and convective form is used to convert water (flowing through the tubes lining the boiler walls) into steam. The high-pressure steam is passed into a turbine, causing the turbine shaft to rotate at high speed. A generator is mounted at one end of the turbine shaft where electricity is generated (via rotating coils in a strong magnetic field). The steam exiting from the turbines condenses, and the condensed water is recycled to the boiler to be used once again. A schematic of a typical PC fired boiler is shown in Figure 1.1.

The PC fired systems were developed in the 1920s. This process brought advantages that included a higher combustion temperature, improved thermal efficiency, and a lower requirement of excess air for combustion. Improvements continue to be made in conventional PC fired power station design, and new combustion technologies are being developed. The focus of considerable efforts by the coal industry has been on improving the efficiency of PC fired power plants. Increasing the efficiency offers several benefits (Burnard and Bhattacharya, 2011):

- Lower coal consumption per megawatt (MW) (i.e., resource preservation)
- Reduced emission of pollutants (e.g., SOx [sulfur oxides], NOx [nitrogen oxides], etc.)

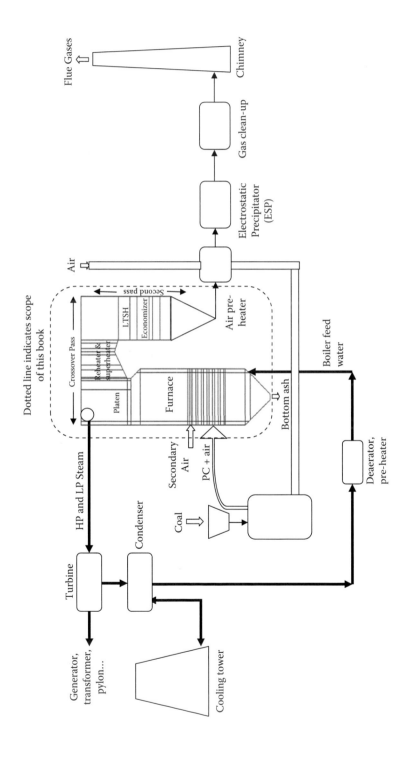

FIGURE 1.1

Schematic of typical pulverized coal fired power plant and scope of this book.

- Reduced CO_2 emission (greenhouse gas): a one percentage-point improvement in overall efficiency can result in ~3% reduction in CO_2 emissions
- Existing power plant efficiencies in developing countries are generally lower. The coal use in these countries for electricity generation is increasing. The improvements in efficiencies of PC fired boilers will therefore have a significant impact on resource conservation and reduction in emissions of CO_2 per megawatt (MW). There is a huge scope for achieving significant efficiency improvements as the existing fleet of power plants is replaced over the next 10 to 20 years.

In recent years, considerable progress has been made in the development of more efficient coal-based systems such as supercritical (SC) and ultra-supercritical (USC) steam cycles. These also include systems based on coal gasification (such as IGCC: integrated gasification combined cycle). However, while there are several proposals for further commercial demonstrations of IGCC and some are being constructed, PC combustion-based plants continue to dominate new plant orders. It may be possible that IGCC will penetrate the market on a large scale only where the co-production of power and chemicals can be economically demonstrated.

Considering the dominance of PC fired systems for electricity generation, several efforts have been made to enhance the overall efficiencies of these systems. The efficiency of a PC fired boiler depends on a variety of factors:

- Plant design (milling, burner, heat recovery systems, and so on)
- Operating conditions (pressure/steam conditions)
- Quality of coal
- Turbine efficiencies
- Ambient conditions
- Operating protocols
- Maintenance practices

In this book we restrict the scope of our discussions to the combustion side of the boiler. The components of the PC fired boiler other than the combustion side are outside the scope of this book. Even for the combustion side, the scope is restricted: from burner/excess air entry points to the exit of flue gases (and fly ash) from the heat recovery section. The milling and conveying of coal particles, as well as further processing of flue gas beyond the heat recovery section (electrostatic precipitators, etc.), are beyond the scope of the present work. The scope of this book is clearly marked over a schematic of a power plant shown in Figure 1.1. Therefore, in the context of this book, PC boiler and PC boiler efficiency essentially refer to the combustion chamber of the PC fired boiler and thermal efficiency, unless stated otherwise.

Several developments on higher-temperature (and -pressure) systems—namely, SC and USC pulverized coal fired boilers—have been realized in recent years (or are being realized). While sub-critical plants achieve overall efficiencies of 32% to 38% (on an LHV [lower heating value] basis), state-of-the-art USC boilers are expected to realize overall efficiencies of 42% to 45% (LHV basis). Single-unit capacities of these plants have reached 1,100 MWe (megawatt electrical). Parallel developments are also being made in oxy-fuel technologies to facilitate CO_2 capture. Efforts have been made to reduce other emissions, such as NOx, SOx, and particulates. Significant research is being done in areas such as new emission control technologies, state-of-the-art boiler designs, and low-temperature heat recovery systems for improved plant efficiencies; systems for improved ash disposability and reduced waste generation; and optimum system integration and plant controls. These research areas are still reinventing themselves.

Key components of efficiency improvement initiatives depend on a better understanding of the burning of coal and subsequent heat transfer for converting water into steam. The overall performance of the boiler critically depends on our understanding as well as our ability to manipulate the following key processes:

- Gas-solid flow (through ducts, burner nozzles, combustion chamber, heat recovery systems, electrostatic precipitators, etc.):
  - Gas: air/flue gas
  - Particles: coal/ash

- Combustion:
  - Volatile combustion
  - Char combustion
  - Pollutant formation and role of furnace aerodynamics

- Radiative and convective heat transfer

Technology advancements in PC fired boilers require a thorough and quantitative understanding of these aspects. Anyone interested in enhancing boiler performance will want to establish quantitative relationships among fuel (coal) characteristics, hardware design of boiler, and operating protocols with boiler efficiency. Such a quantitative relationship can then be used to optimize fuel mix and design as well as operating practices. Computational modeling, particularly computational flow modeling (CFM), plays a crucial role in developing such an understanding as does the ability to tailor the design and operating protocol to enhance overall performance.

Over the years, computational fluid dynamics (CFD) tools have evolved as powerful design and predictive tools to simulate complex equipment (Ranade, 2002). Computational fluid dynamics (CFD) is a body of knowledge and techniques to solve mathematical models (primarily based on conservation of mass, momentum, and energy) on digital computers. The development of high-performance computers, advances in the physics of fluids, and advances in numerical techniques and algorithms has made it increasingly possible to use CFD models for complex reacting systems such as PC fired boilers. It is, however, necessary to adapt CFD techniques and develop an appropriate CFM approach for meaningfully applying them to complex systems like PC fired boilers.

This book is written with the intention of assisting practicing engineers and researchers in developing such an approach. It describes the methodology to formulate a CFD model for simulating PC fired boilers. Although most of the results included here are for an old-generation 210-MWe unit, the methodology can be extended in a straightforward way to model advanced supercritical, ultra-supercritical, and oxy-fuel combustion systems.

CFD models/simulations of PC fired boilers are usually computationally expensive and time consuming. Therefore, even if these models/simulations provide extensive insight into processes occurring in PC fired boilers, they are not very suitable for real-time optimization (which requires much quicker response times). A complementary approach called reactor network modeling (RNM) is therefore developed and presented in this book. The RNMs are essentially lower-order computational models (typically with a few hundred variables, in contrast to a few million or more variables used in CFD models) that can be used for overall optimization and even for real-time process control. The methodology and process of developing an RNM based on data extracted from detailed CFD models of PC fired boilers are discussed here. Efforts are made to provide guidelines for

- Characterizing coal using a drop-tube furnace and a thermogravimetric analyzer (TGA)
- CFD modeling of a PC fired boiler
- Developing reactor network models (RNMs)
- Applying computational models to industrial practice

The presented results and discussion will also provide insight into the complex processes occurring in PC fired boilers. The basic information and key issues of coal fired boilers are discussed in Section 1.1. Key aspects of CFM with reference to potential application to coal fired boilers are then discussed in Section 1.2, and the organization of the book is briefly discussed in Section 1.3.

1.1 Coal Fired Boilers

Worldwide, coal fired generating capacity is expected to reach approximately 2,500 GW (gigawatt) by the end of 2020. This is an increase of nearly 60% since 2008, and more than 55% of the projected new generating capacity is expected to be in Asia (Mcllvaine, 2009). The capacity enhancement is based on the addition of advanced low-emission boiler systems. Continuing efforts are also being made on existing power plants to achieve higher efficiency, reliability, and availability with low maintenance, while complying with stringent emissions regulations for CO_2, SOx, NOx, and particulates. Coal fired boilers have undergone major innovations in order to satisfy both economics and increasing stringent environmental regulations. In principle, solid-fuel combustion technologies are divided into three categories:

1. Bed or grate combustion
2. Suspension or pulverized combustion
3. Fluidized bed combustion

The traveling grate stoker was the early coal combustion system for power generation. Traveling grate stokers are capable of burning coals over a wide range of coal rankings (from anthracite to lignite). The typical particle sizes are 1 to 5 cm (centimeter) with residence times of 3,000 to 5,000 s (second). The flame temperature is around 1,750K (Essenhigh, 1981). Stoker firing is not able to scale beyond 50-MWe unit capacities. The boiler efficiency gets suppressed by the high excess air (about 40%) that was required for acceptable coal burnout.

In 1946, Babcock and Wilcox introduced the cyclone furnace for use with slagging coal (i.e., coal that contains inorganic constituents that will form a liquid ash at temperatures of ~1,700K or lower); this was the most significant advance in coal firing since the introduction of pulverized coal firing (Miller, 2005). A cyclone furnace provides the benefits obtained with PC firing and also has the advantages of utilizing slagging coals and reducing costs due to less fuel preparation (i.e., fuel can be coarser and does not need to be pulverized). In slagging combustion, the boiler tubes in the lower part of the furnace are covered by a refractory to reduce heat extraction and to allow the combustion temperature to rise beyond the melting point of the ash. The temperature must be sufficiently high for the viscosity of the slag to be reduced to about 150 Poise, which is necessary for removal in liquid form. The most notable application of slagging combustion technology in the United States is the cyclone furnace, in which about 85% of the coal ash can be removed in molten form in a single pass without ash recirculation. Because of the high temperature and the oxidizing atmosphere, slagging furnaces produce very high NOx emissions and therefore fell into disfavor in the 1970s.

Fluidized bed boilers for utilizing coal were originally developed in the 1960s and 1970s; they offer several inherent advantages over conventional combustion systems, including the ability to burn coal cleanly by reducing SO_x emissions during combustion (i.e., *in situ* sulfur capture) and generating lower emissions of NO_x. In addition, fluidized bed boilers provide fuel flexibility as a range of low-grade fuels can be burned efficiently. In fluidized bed combustion (FBC), crushed coal of 5 to 10 mm (millimeter) is burned in a hot fluidized bed of 0.5- to 3.0-mm-sized inert solids. The typical particle residence time is around 100 to 500 s (Essenhigh, 1981). Less than 2% of the bed material is coal; the rest is coal ash and limestone, or dolomite, which are added to capture sulfur during the course of combustion. The bed is cooled by steam-generating tubes immersed in the bed to a temperature in the range of 1,050K to 1,170K (Baranski, 2002). This prevents the softening of the coal ash and the decomposition of $CaSO_4$, which is the product of sulfur capture. The heated bed material, after coming into direct contact with the tubes (heating by conduction), helps improve the efficiency of heat transfer. Because this allows coal plants to burn at lower temperatures, less NOx is emitted. This technology, however, was limited to small industrial-sized fluidized boilers and was not useful for very large steam capacities like those of PC fired units unless use of high pressures.

Circulating fluidized bed (CFB) boilers were developed to enhance the flexibility of using a wider range of fuels. CFB boilers involve a circulating fluidized bed of inert material with a hot cyclone to recirculate solids particles. Usually, combustion in CFB boilers takes place at temperatures from 1,075K to 1,175K, resulting in reduced NOx formation compared with conventional PC fired boilers. Another advantage of CFB combustors is their ability to handle low calorific value fuels, *in situ* sulfur capture, no need to pulverize the fuel, and comparatively lower capital costs than PC boilers. The thermal efficiency of CFB boiler units is lower by 3 to 4 percentage points than that for equivalent-sized PC boilers (Western Governors' Association, 2008). CFB boilers represent the market for medium-scale units (typically < 300 MWe) in terms of utility requirements. They are used more extensively by industrial and commercial operators in small to medium range sizes (50 to 100 MWe), both for the production of process heat and for on-site power supply.

Pulverized coal (PC) combustion became a widely accepted combustion system for power generation in the period between 1900 and 1920. This was the major development in order to take advantage of the following:

- Higher volumetric heat release rates of pulverized coal (MW/m^3)
- Increased system efficiencies using superheaters (heat-exchange surface to increase the steam temperature), economizer (heat-exchange surface to preheat the boiler feed water), and combustion air preheaters (heat-exchange surface to preheat combustion air)
- Improved material of construction, which will allow for higher pressures in the steam generator (>83 barg [bar gauge])

The development of the superheater, reheater, economizer, and air pre-heater played a significant role in improving overall system efficiency by absorbing most of the heat generated from burning the coal. The separation of steam from water and the use of super-heaters and reheaters allowed for higher boiler pressure and larger capacities. Hence, the pulverized coal combustion system became widespread due to increased boiler capacity, and improved combustion and boiler efficiencies over stoker-fired boilers. The PC fired boiler was found to be the most suitable for utility power plants and contributes to more than 50% of the world's electricity demand.

Advances in materials of construction, system design, and fuel firing have led to increased capacity and higher steam operating temperatures and pressures. There are two basic PC-fired water tube steam generators: (1) subcritical drum-type boilers with nominal operating pressures of either 131 or 179 barg, and (2) once-through supercritical units operating at 262 barg. These units typically range between 300 and 800 MW (i.e., produce steam in the range of 900 to 3,000 tons/hour). Ultra-supercritical units entered into service in 1988. These units operate at steam pressures of 310 barg and steam temperatures of 838K, with capacities as high as 1,300 MW. This book, as mentioned earlier, focuses on developing a better understanding of the various processes occurring in the most widely adopted PC fired boilers. Some aspects of these tangentially PC boilers are discussed here.

A typical tangentially fired PC boiler is shown in Figure 1.2(a). There are three major parts of the boiler: furnace, crossover pass, and second pass. In a tangentially fired furnace, the burners generate a rotational flow in the furnace by directing the jets tangent to an imaginary circle whose center is located at the center of the furnace. The resultant swirling and combusting flow generates a fireball at the center of the furnace, where the majority of combustion occurs (Figure 1.2(b)).

In PC combustion, the coal is dried and crushed into fine particles, and the powdered coal is pneumatically transported to the burners. Typical particle sizes used in PC boilers range from 10 to 100 μm (micrometer). The design of the combustion chamber must provide sufficient residence time for the coal particle to complete its combustion, and for the fly ash to cool down to below its "softening temperature" to prevent the build-up of ash deposits on heat exchanger surfaces. A typical residence time of particles in PC boilers is approximately 10 s (Essenhigh, 1981). The transport air that carries the coal from the mill to the burners must be maintained at around 373K to avoid ignition hazards. This transport air (also called fuel air [FA]) is about 40% to 50% of the total air injected into the boiler. The remaining air other than the FA enters the PC boiler via secondary air (SA) inlets and over fire air (OFA). The air entering through these other inlets is usually preheated to 550K.

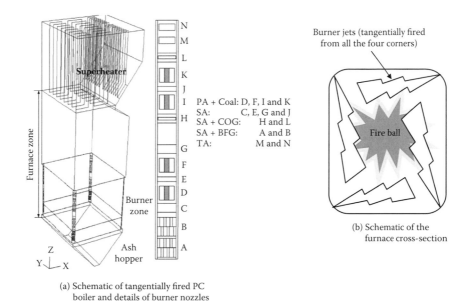

FIGURE 1.2

Tangentially fired furnace of pulverized coal fired boiler. (Figure 1.2 (a) reprinted with permission from Fang, Q., Musa, A.A.B., Wei, Y., Luo, Z. and Zhou, H., Numerical simulation of multi-fuel combustion in a 200 MW tangentially fired utility boiler, *Energy & Fuels*, 26, 313–323. (2012); © 2012 American Chemical Society.)

The design of the furnace is influenced by the type of burner system adopted in the design, such as a wall-mounted circular type or a corner-fired rectangular shape. Circular burners are usually positioned perpendicular to the combustion chamber walls, while the vertical nozzle arrays are in the corners, firing tangentially to the circumference of an imaginary cylinder in the middle of the combustion chamber. Wall-mounted burners can also be designed to provide tangential entry to the coal and air, forming a rotating fireball in the center of the furnace. The majority of boilers in countries such as India are tangentially fired and use rectangular slotted burners. The pulverized coal, air mixture (primary air), and secondary air are injected from sixteen to twenty-four burners located in the four corners of the furnace (Figure 1.2(b)).

The heat generated due to the combustion reactions is radiated and transported to the water walls of the boiler to generate steam. This steam is further superheated in superheaters at the crossover pass by radiative and convective heat transfer. The last few percentages of residual carbon in the char burns in an environment of depleted oxygen concentration and reduced temperature before the fly ash leaves the combustion chamber and enters the pass of convective heat exchangers. In the majority of cases, most of the fly ash formed in PC combustion is removed from the flue gas in the form of dry

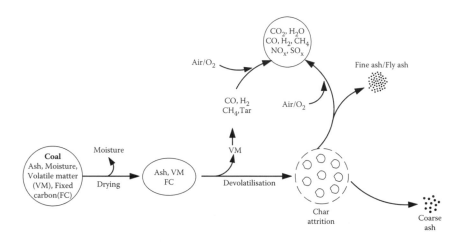

FIGURE 1.3
Particle-level processes during coal combustion.

particulate matter, with a small portion (about 10%) of the coal ash falling off the tube walls as semi-molten agglomerated ash, which is collected from the bottom hopper of the combustion chamber ("bottom ash"). Pulverized coal fired boilers offer high combustion intensities ($\sim$1 MW/m^3) and high heat transfer rates ($\sim$1 MW/m^2).

The coal particles injected in PC fired boilers get heated. Moisture and volatile materials are released during the drying and devolatilization stage. The char content of the particles eventually burns, and the remaining ash can exit the boiler in the form of fly ash (finer particles) or bottom ash (coarser particles). These processes are schematically shown in Figure 1.3. Pyrolysis or devolatilization results in a vast array of products such as CO, tar, H_2, H_2O, HCN, hydrocarbon and gases, etc. These products react with oxygen in the vicinity of the char particles, depleting oxygen and increasing the temperature. These complex reaction processes are very important in controlling the nitrogen oxides, formation of soot, stability of coal flames, and ignition of char. The char combustion may then occur along with the combustion of the volatile species in the gas phase. After the completion of char combustion, the inorganic constituents remain as ash. Typically, ash contains SiO_2, Al_2O_3, Fe_2O_3, TiO_2, CaO, MgO, Na_2O, K_2O, SO_3, and P_2O_5 (primarily the first 3 compounds). Ash composition varies with the rank of the coal. The ash material may further decompose to form slag at high temperatures (beyond the ash fusion temperature). Nitrogen is released from the coal and forms nitrogen oxides (NOx). The sulfur released during coal combustion forms sulfur oxides (SOx).

The process of char combustion is of central importance in industrial PC fired boiler applications. However, despite extensive investigations over the past half-century, the mechanism of the char/oxygen reaction is not completely understood because of a number of factors, including the reaction due to pore growth and mass transfer effects (Williams et al., 2001). It is further complicated by the influence of particle size distribution, char mineral content, and fragmentation of the char particle. The rate-limiting step in the combustion of a char particle can be chemical reactions, or the diffusion of gases into the particle, or a combination of these factors.

The furnace is a key part of the boiler where turbulent mixing, two-phase flow, exothermic reactions, and radiative heat transfer dominate. Preliminary calculations shows that combustion of a typical 70 µm coal particle require nearly 1 s at typical furnace temperatures. The industrial data, however, confirm the presence of unburned carbon in fly ash (~5%) as well as in bottom ash, which affects the thermal efficiency of the boiler and the salability (value) of the ash. Turbulent and recirculating flow affects the mixing of the fuel and oxidants, and thereby the rate of the combustion reaction. Understanding the behavior of coal fired boilers and translating this understanding into computational models for simulating such boilers is complicated by the existence of turbulent, multi-phase recirculating flows, coupled with chemical reactions and radiative heat transfer. The existence of a wide range of spatiotemporal scales (chemical reactions occurring on molecular scales to micron-size particles to tens of meters of boiler; reaction time scales on the order of microseconds to particle residence time on the order of a few

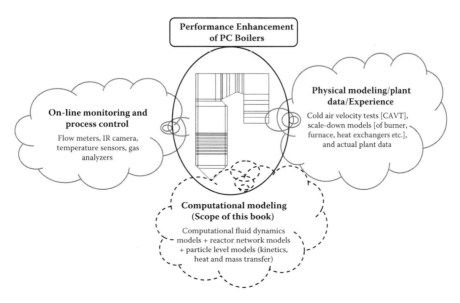

FIGURE 1.4
Approaches for performance enhancement of boilers.

seconds) often complicates the matter further. The thermo-fluid interaction processes between neighboring burners and between the burners and the furnace as a whole are complex and not yet well understood.

As mentioned earlier, significant efforts have been made and are still being made to enhance the performance of PC fired boilers. Possible ways of enhancing the performance of such boilers (see Figure 1.4) may be broadly grouped into two types:

1. On-line monitoring and process control. These methods are usually based on data-driven models and can be effectively used for optimizing the operation of a PC fired boiler over a narrow range of parameters.

2. Establishing the relationship between performance and hardware/ operating protocols. These methods may be further classified into two subtypes:
    - Based on physical modeling, data and empirical correlations, experience, and heuristic relationships
    - Phenomenological modeling: representing the behavior of a PC fired boiler using conservation equations of mass, momentum, and energy

In this book we mainly restrict our scope to the discussion of phenomenological models. Considering the complexity of PC fired boilers, most phenomenological models of PC boilers require the help of a digital computer for their solution. The combined set of underlying phenomenological models, initial and boundary conditions, numerical methods, and their computer implementation is broadly called a "computational model." This book essentially deals with the development and application of such computational models for simulations and for performance enhancement of PC fired boilers.

The performance of PC fired boilers operating with high-ash coal (like Indian coal) faces some special challenges. These include problems of ash handling, slagging, ash deposition on heat exchanger surfaces, as well as unburned char in the fly ash. With emission norms becoming more stringent and increasing pressures on enhancing overall efficiency of PC fired boilers, it is essential to develop relationships among the availability, performance, and reliability of these boilers with factors such as burner design, operational protocols, coal quality, age of boiler, furnace hardware configuration, etc.

Computational modeling offers an excellent way to develop such relationships and to develop an ability to optimize boiler performance. Some aspects of computational flow modeling are discussed in Section 1.2.

1.2 Computational Modeling

Computational models attempt to represent processes occurring in PC fired boilers with the help of conservation equations of mass, momentum, and energy. A generic conservation equation for any conserved quantity ϕ can be represented as

$$
\begin{array}{lllll}
\text{accumulation of} & \text{rate of change of} & & \text{rate of change of} & & \text{rate of change of} \\
\text{component } \phi & = & \text{component } \phi \text{ due} & + & \text{component } \phi \text{ due} & + & \text{component } \phi \text{ due to} \\
& & \text{to convection} & & \text{to dispersion} & & \text{reaction / mass transfer}
\end{array}
$$

$$(1.1)$$

If there are any additional modes of transport for ϕ, corresponding additional terms may appear on the right-hand side of Equation (1.1). This general balance equation can be mathematically represented in a variety of forms, depending on various underlying assumptions. One can use a classical chemical reaction engineering (CRE) approach and develop a RNM to represent the PC fired boiler. Alternatively, one can use a CFD-based framework to represent the PC fired boiler. The term "computational modeling" essentially covers the selection of a modeling approach, development of model equations and boundary conditions, development of numerical methods to solve model equations, mapping of numerical methods on digital computers to generate solutions of model equations, verification of developed computer implementation and validation of the underlying phenomenological model, carrying out simulations to gain a better understanding, and using that gained understanding to enhance performance in practice.

Computational modeling of PC fired boilers therefore requires broad-based expertise in the various aspects of PC combustion and an in-depth understanding of various aspects of computational modeling (CFD and CRE), along with a generous dose of creativity. Ranade (2002) has discussed the application of computational flow modeling for a variety of chemical reactors. The discussion is very relevant for modeling and simulating the PC fired boiler as well and is therefore recommended for further reading. Computational modeling requires relatively few restrictive assumptions, can handle complex configurations, and can incorporate a variety of processes simultaneously. Simulations based on such computational models often serve as a bridge between theory and reality. Computational models allow switching on and off the various interactions included in the model in order to understand the relative contributions of each individual process, which is extremely difficult—if not impossible—to achieve in experiments. These advantages of computational models are conditional and may be realized only when the model equations capture key underlying processes and are solved accurately. This is easier said than done, especially for complex equipment like the PC fired boiler. It must be remembered that simulated results based on computational models will always be approximate. There can be various reasons for differences between computed results and "reality." Errors are likely to arise from each part of the process used to generate computational simulations:

- Model equations
- Input data and boundary conditions
- Numerical methods and convergence
- Computational constraints
- Interpretation of results, etc.

It is indeed necessary to develop an appropriate methodology to harness the potential of computational modeling for engineering analysis and design despite some of the limitations. In this book we discuss such a methodology for applying computational modeling to PC fired boilers.

Recent advances in understanding multiphase flows, turbulence, and combustion, coupled with advances in computing power and numerical methods, provide an opportunity to develop a better understanding and better computational models for simulating coal fired boilers. The development of advanced industrial burners, furnaces, and boilers with the improved performance of higher efficiency and lower pollutant emissions is the major goal of combustion researchers, furnace designers, and manufacturers. To realize the goal, on the one hand, the new technical concepts and novelties for different combustion routes and processes must be continuously developed. On the other hand, more efficient and economic tools, such as computer simulation using CFD technology, are also extremely important to develop and apply to the design process and performance simulation of the new, more advanced furnaces.

In recent decades, compared with traditional experimental methods and physical modeling methods, computer simulation with numerical methods is considered a more attractive effective design tool. With recent advances in our understanding of the physics of flows and the availability of large and powerful computational resources, comprehensive computer simulation has become possible even for very complicated applications in industry. The influence of burner design on flame stability and temperature distribution within the furnace, and hence its effect on overall performance of the boiler, can now be investigated by state-of-the-art CFD technologies (Park et al., 2013; Modelinski, 2010; Choi and Kim, 2009; Belosevic et al., 2006; Pallares et al., 2005; Yin et al., 2002; Fan et al., 2001; Eaton et al., 1999; Boyd and Kent, 1986).

The development and application of comprehensive, multidimensional computational combustion models are increasing at a significant pace all over the world. These combustion models are becoming more readily accessible as features in commercially available CFD computer codes. A number of commercial CFD codes have been developed and are used in practice. Ranade (2002) has listed some of the commonly used CFD codes and tools. Updated information can be found on the websites of the various vendors. There are also open-source CFD codes available to potential users. A partial listing of available CFD resources can be found at www.cfd-online.com. Simulations made with such computer codes offer great potential for use in analyzing, designing, retrofitting, and optimizing the performance of fossil-fuel combustion and conversion systems. However, CFD is expensive in terms of computational time and resources. A few researchers have tried in the past to develop phenomenological models to simulate PC fired boilers where key information about the sizing of zones and fluid flow is extracted from CFD simulations (Diez et al., 2005; Falcitelli et al., 2002). This is a very useful approach as these models require orders of magnitude lower computational resources and thereby can be integrated with the on-line optimization and process control platforms. An attempt is made in this book to provide adequate information to the reader for developing an RNM for a PC fired boiler. Discussion on applying computational models to practice is included in a separate chapter. The details of the organization of this book are briefly discussed in Section 1.3.

1.3 Organization of the Book

This book is organized to facilitate the development of modeling tools comprised of computational fluid dynamics (CFD) and phenomenological models to simulate various processes occurring in typical PC fired boilers. Most of the presented results are for the typical 210-MWe tangentially fired PC boiler. All the key aspects necessary to establish computational models such as the following are included:

- Devolatilization and char combustion of high-ash Indian coal
- Models to interpret drop-tube furnace data
- Development of CFD models
- Influence of operating conditions on performance of 210-MWe boilers
- Development of phenomenological/reactor network models (RNMs)

A chapter-wise outline of the book is provided below:

- Chapter 1 introduces the topic and provides the background, motivation, and scope of the book.
- Chapter 2 discusses the approach toward the computational modeling of PC boilers. The role of computational models is discussed and a brief review of different modeling approaches is provided. Key

issues while applying computational models in practice and some tips for facilitating their application are highlighted.

- Chapter 3 focuses on the characterization of coal based on thermo-gravimetric experiments (TGA), as well as the drop-tube furnace (DTF). The chapter emphasizes the importance and use of 2D axi-symmetric CFD models over conventional 1D models in estimating kinetic parameters.

- Chapter 4 provides a systematic approach for the development of a detailed CFD model for tangentially fired pulverized coal boilers. A brief review of computational modeling of PC boilers is included. The methodology and presented results will be useful in understanding gas flow, particle trajectories, the extent of char burnout, gas temperature, and species concentration field within a typical boiler. The crossover pass characteristics (uneven distribution of flow and temperature) of tangentially fired boilers and other key issues are discussed with the help of the computational model. The utility of the developed model to quantify the sensitivity of boiler performance on operating parameters is demonstrated. And the influence of parameters that can be varied during boiler operation, such as excess air and burner tilt on boiler performance, is discussed. A sample of simulated results is presented in an effort to understand boiler performance when high ash content Indian coal is blended with low ash and highly volatile lignite coal in various ratios.

- Chapter 5 discusses the development of a state-of-the-art reactor network model based on off-line interaction with the analysis of the results obtained with the CFD model. This model framework translates the information gained from the detailed CFD model to a readily usable engineering scale model for actual plant implementation. The model is based on the reactor network (RN) approach, comprised of networked multiple zones representing key sections of the boiler. The positioning and size of different zones depend on the underlying fluid dynamics. The effect of key operating protocols such as burner tilt is accounted for through appropriate correlations developed from CFD simulations. Application of the model is illustrated for a typical 210-MWe PC fired boiler. The results obtained with the RNM are compared with the overall CFD results.

- Chapter 6 summarizes the work and results presented in Chapters 2 through 5. It also discusses judicious use of mathematical models in practice, along with common pitfalls, and provides guidelines for the effective use and enhanced utility of mathematical models.

- Chapter 7 provides an overall summary and key conclusions. Some comments and suggestions for possible extensions and the path forward are also included in this final chapter.

References

Baek, S.H., Park, H.Y., and Ko, S.H. (2014). The effect of the coal blending method in a coal fired boiler on carbon in ash and NOx emission, *Fuel*, 128, 62–70.

Baranski, J. (2002). Physical and Numerical Modeling of Flow Pattern and Combustion Process in Pulverized Fuel Fired Boiler, Ph.D. thesis, Royal Institute of Technology, Sweden.

Belosevic, S., Sijercic, M., Oka, S., and Tucakovic, D. (2006). Three-dimensional modeling of utility boiler pulverized coal tangentially fired furnace, *Int. J. Heat and Mass Transfer*, 49, 3371–3378.

Boyd, R.K. and Kent, J.H. (1986). Three-dimensional furnace computer modeling, In *21st Symp. (Int.) on Combustion,* The Combustion Institute, pp. 265–274.

Burnard, K. and Bhattacharya, S. (2011). Power Generation from Coal: Ongoing Developments and Outlook, OECD/International Energy Agency (IEA) (www.iea.org).

Choi, C.R. and Kim, C.N. (2009). Numerical investigation on the flow, combustion and NOx emission characteristics in a 500 MWe tangentially fired pulverized-coal boiler, *Fuel*, 88 (9), 1720–1731.

Díez, L.I., Cortés, C., and Campo, A. (2005). Modelling of pulverized coal boilers: Review and validation of on-line simulation techniques, *Appl. Thermal Eng.*, 25 (10), 1516–1533.

Eaton, A.M., Smoot, L.D., Hill, S.C., and Eatough, C.N. (1999). Components, formulations, solutions, evaluation, and application of comprehensive combustion models, *Progr. Energy and Combustion Sci.*, 25, 387–436.

Essenhigh, R.H. (1981). In Lowry, H.H., Ed., *Chemistry of Coal Utilization, 2nd volume.* Wiley, New York.

Falcitelli, M., Tognotti, L., and Pasini, S. (2002). An algorithm for extracting chemical reactor network models from CFD simulation of industrial combustion systems, *Combustion Sci. Technol.*, 174, 27–42.

Fan, J.R., Zha, X.D., and Cen, K.F. (2001). Study on coal combustion characteristics in a w-shaped boiler furnace, *Fuel*, 80, 373–381.

Fang, Q., Musa, A.A.B., Wei, Y., Luo, Z., and Zhou, H. (2012). Numerical simulation of multi-fuel combustion in a 200 MW tangentially fired utility boiler, *Energy & Fuels*, 26, 313–323.

IEA (2010a). Energy Technology Perspectives: Scenarios & Strategies to 2050, OECD/IEA, Paris.

IEA (2010b). World Energy Outlook 2010, OECD/IEA, Paris.

Mcllvaine (2009). Eastern market shows promise, *World Pumps*, 153, 18–20.

Miller, B.G. (2005). *Coal Energy Systems,* Elsevier Academic Press., Burlington, MA, USA.

Modelinski, N. (2010). Computational modeling of a utility boiler tangentially-fired furnace retrofitted with swirl burners, *Fuel Processing Technol.*, 91, 1601–1608.

Pallares, J., Arauzo, I., and Diez, L.I. (2005). Numerical prediction of unburned carbon levels in large pulverized coal utility boilers, *Fuel*, 84, 2364–2371.

Park, H.Y., Baek, S.H., Kim, Y.J., Kim, T.H., Kang, D.S., and Kim, D.W. (2013). Numerical and experimental investigations on the gas temperature deviation in a large scale, advanced low NOx, tangentially fired pulverized coal boiler, *Fuel*, 104, 641–646.

Ranade, V.V. (2002). *Computational Flow Modeling for Chemical Reactor Engineering*, Academic Press, London.

Western Governors' Association (2008), Deploying Near-Zero Technology for Coal: A Path Forward (http://www.westgov.org/images/dmdocuments/zero-coal08. pdf, accessed May 1, 2014).

Williams, A., Pourkashanian, M., and Jones, J.M. (2001). Combustion of pulverised coal and biomass, *Progr. Energy and Combustion Sci.*, 27, 587–610.

Yin, C., Caillat, S., Harion, J., Baudoin, B., and Perez, E. (2002). Investigation of the flow, combustion, heat-transfer and emissions from a 609 MW utility tangentially fired pulverized-coal, *Fuel*, 81 (8), 997–1006.

2

Toward the Computational Modeling
of Pulverized Coal Fired Boilers

A pulverized coal (PC) fired boiler is a complex chemical reactor consisting of many complex processes, such as gas-solid flows, combustion, heat transfer (conduction, convection, and radiation) and phase change, all in one piece of equipment. Key features of PC fired boilers were discussed in Chapter 1. In order to design and operate PC fired boilers in the best possible way, several design variables and operating parameters must be optimized. Some of these include

- Size and configuration of the boiler
- Design, number, and location of the burners
- Characteristics of the fuel (coal)
- Particle size/size distribution of the pulverized fuel
- Extent of excess air to be used
- Feed temperature (of air and coal particles)

Design and operating engineers are interested in maximizing the thermal efficiency of boilers by eliminating or minimizing unburned carbon in fly ash and bottom ash. Internal heat exchangers and the fluid dynamics of the furnace must be designed to maximize efficiency on the one hand and minimize downtime of the boiler on the other hand (often required for cleaning deposits on internal heat exchangers and water walls of the boiler). Through years of design and operating experience, significant expertise and guidelines have been developed by practicing engineers. However, as discussed in Chapter 1, it is always desirable to complement this accumulated experience with computational models and simulation tools. Computational models allow design and operating engineers to gain better understanding and carry out virtual experiments to evaluate new and innovative ideas. The inherent complexities of PC fired boilers, however, make the task of computational modeling these boilers quite complex and nontrivial. This book essentially discusses the details of computational modeling of PC fired boilers. This chapter discusses the overall approach toward developing computational models of PC fired boilers.

The first step toward developing "tractable" computational models of a complex system such as a PC fired boiler is to develop an appropriate approach. The following two guidelines should be considered while developing such an approach:

1. Occam's razor: It is futile to do with more what can be done with less.
   a. Simple/conventional models may provide an adequate solution. In such a case, there is no point in using more complex models without corresponding gains in accuracy or reliability.
   b. Often, the combination of multiple models works better than a single monolithic model.
2. One should always try to make things as simple as possible but not simpler (sentence attributed to Einstein).
   a. Simplifications are essential for developing tractable models of real-life processes. However, it is important to understand and to identify the difference between making things simple and simpler. If key controlling subprocesses are not considered adequately while modeling, it may defeat the purpose of modeling.
   b. It is therefore essential to match the complexity of the problem with the analysis tool. The art of modeling is essentially understanding the differences between "simple" and "simpler."

Anyone interested in attempting the computational modeling of PC fired boilers should be familiar with some basic concepts in the mathematical modeling of physical processes (see, for example, books by Denn, 1986; Aris, 1978; and Polya, 1967). The relative importance and roles of governing equations, constitutive equations, boundary conditions, and input data must be clearly understood while interpreting the results and drawing engineering conclusions based on the simulated results. It is also important to recognize the possibility of employing a hierarchy of models to develop the necessary understanding and to obtain the required information for achieving complex engineering objectives.

Ranade (2002) has succinctly discussed the need for using a hierarchy of models with an analogy to the variety of vehicles available for transport that is reproduced here. Various alternative vehicles—ranging from bicycle, scooter, car, and helicopter to aircraft—are available for a person who wants to travel. Each of these vehicles has unique features and a corresponding range of applications. The availability of powerful alternatives for transport has not made other, less powerful modes obsolete. More often than not, the best way to travel the desired distance is based on using different vehicles for different parts of the journey. Similarly, there will be a hierarchy of mathematical models, each having some unique features with a corresponding range of applications that may be used to construct as complete a picture of the physical process as possible. Computational flow modeling (CFM) is certainly a very powerful tool; and, in principle, a self-content, comprehensive mathematical model can be constructed to simulate the behavior of industrial PC boilers within the CFM framework. However, it may be inefficient to use such a complex model to obtain certain information that might be obtained by relatively simple models.

The overall approach based on using a hierarchy of models is advocated in this book and briefly discussed in Section 2.1. Ranade (2002) has discussed previously some aspects of the approach.

2.1 Overall Approach

The first essential step while developing an overall approach toward the computational modeling of PC boilers (or, for that matter, any other real-life process) is to clearly identify and define the goals for modeling. This appears to be quite obvious. Unfortunately, in practice, the causes of many not-so-successful performance enhancement projects may be traced to inadequate attention to this step. The goals of computational modeling are usually linked to the performance of the boiler. It is therefore necessary to clearly spell out the key performance parameters or "wish list" for a PC fired boiler. This wish list about the overall performance of the boiler may contain all the items by which boiler performance will be judged (some of it may also be just qualitative). This is discussed again in Chapter 6.

The next step is to identify key processes occurring in a PC fired boiler that control its performance. It is important to relate these key processes to underlying transport processes. This will help to identify relevant transport processes with reference to the objectives defined in the "wish list" of key performance parameters. Example of these two lists in the context of PC fired boilers can be found in Chapter 1. This process will also help evolve specific goals for and clearly define expectations from the computational model. Generic expectations from the computational model include

- Detailed analysis at the early stages in the design cycle for less money, less risk, and less time
- The capability to make *a priori* predictions of process performance with just the knowledge of geometry and operating parameters
- Prediction of what could or would happen as a result of a specific design, thereby steering the design in promising directions

These generic expectations must be translated into specific modeling goals. At this juncture, it is important to clearly understand the characteristics of the "learning" and "simulation" or "design" models. "Learning" models are developed to understand basic concepts and to obtain specific information about some unknown process(es). The results obtainable from such models may not directly lead to design information but are generally useful in making appropriate engineering decisions. "Design" or "simulation" models, on the other hand, yield information or results that can be directly used for design and engineering. It may be possible to develop a rigorous computational model of a simplified case of a PC fired boiler that can be used as a "learning" tool. For example, complex interactions of turbulence, combustion, and gas-solid flows can be simulated rigorously using the direct numerical simulation (DNS) approach. These simulations can provide a wealth of information about such interactions and can provide data that are very difficult to obtain from experiments. Therefore, although such simulations are restricted to low Reynolds number flows and simple geometry, they can lead to a better understanding of and insight into the development of better phenomenological models. The objectives for developing such a model, obviously, will be very different from that developed for simulating the realistic behavior of industrial PC fired boilers. For the latter case, the computational model must relate the boiler hardware and operating protocol with the performance using finite computational resources. Several such examples can be given for the case of PC boilers as the underlying physics is very complex. A clear understanding and visualization of the expected results and their proposed use is, therefore, essential for defining the objectives of the computational model.

After establishing specific goals, it is necessary to select an appropriate level of complexity of the computational models to meet these objectives. This also includes various other aspects such as the required degree of accuracy of predicted results and available resources. It is essential to give sufficient thought to the required degree of accuracy of the predictions of the computational model right at the first stage. The intended use of the results of the computational model dictates the required accuracy levels. The available time, expertise, and computational facilities also should be examined in order to define realistic goals. Thorough analysis of these issues will allow one to clearly define the goals of the computational model and, in the process, also evolve a methodology for using the results of the flow model to achieve the performance objectives.

It will be useful to make a few comments about the validation of the simulated results and their use for enhancing the performance of industrial PC fired boilers. Even before the validation, it is necessary to carry out a systematic error analysis of the generated computer simulations. The influence of numerical issues on predicted results and errors in integral balances must be checked and ensured to be within acceptable tolerance levels. The simulated results must be examined and analyzed using available post-processing tools. The results must be checked to verify whether the model has captured major qualitative features such as the fireball, temperature imbalances, etc.

Whenever various phenomenological models are used, further quantitative validation of the simulated results is essential. Even if the objective of the computational model is to qualitatively screen possible alternative configurations, it is important to validate the simulations quantitatively to ensure that they have adequately captured the basic phenomena controlling the performance. In many cases, however, the data of industrial PC fired boilers are not available and are difficult to measure. Direct quantitative validation is not possible in such cases. Indirect validation must be carried out for such cases. Often it may be necessary to independently validate various submodels and phenomenology incorporated in the overall computational model. Here again, emphasis should be on verifying whether or not the key processes are adequately simulated. Prior experience and engineering judgment should be used while interpreting simulation results. More discussion on this is included in Chapter 6.

Despite some of these limitations, computational modeling can prove to be a great help in realizing the "wish lists" of the overall performance enhancement of industrial PC boilers. Typically, it is useful to develop some design models to estimate boiler sizing and to evolve preliminary boiler configurations. Several "learning" models can then be developed to understand the various key issues. The understanding gained by the development and application of these "learning" models will be helpful in identifying the need and methodology for developing more sophisticated simulation models to establish the desired optimal boiler design. Different models/steps of the overall approach may be listed as

- Engineering design models
- Single-particle models (see also Chapter 3)
- Boiler-level models
  - Reaction engineering models (see also Chapter 5)
  - Computational fluid dynamics (CFD) models (see also Chapter 4)
- Methodology of applying mathematical models to practice (see also Chapter 6)

The interaction of these steps is shown pictographically in Figure 2.1. The steps in this overall approach are briefly outlined in the following subsections.

2.2 Engineering Design Models

In simple terms, a boiler is an enclosed vessel that provides a means for the heat of combustion to be transferred to water until it becomes heated water or steam. Several design handbooks are available that provide overall

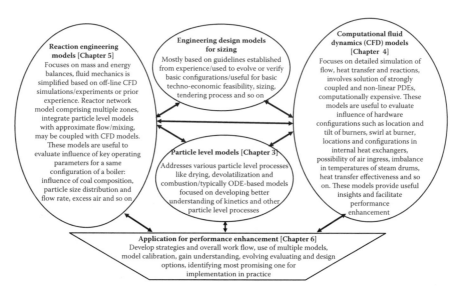

FIGURE 2.1
Modeling of pulverized coal fired boilers.

guidelines for sizing and configuring the boiler (Speight, 2013; Heselton, 2005). Table 2.1 provides a sample of such guidelines.

In principle, it is possible to develop simplified models for understanding overall behavior. However, as decades of experience have accumulated on boiler design and operation, more often than not the guidelines established based on experience are used for sizing and finalizing the overall configuration of the boiler instead of using computational models. These guidelines and sizing models are not discussed in this book. Once the basic sizing and configuration of a PC fired boiler has evolved, more sophisticated models are then used to simulate the behavior of a developed boiler configuration. These models, as discussed previously, are typically used in parallel fashion (see Figure 2.1), and this book essentially focuses on these types of models. Some details of these models are briefly discussed in the following section to illustrate the overall approach.

2.3　Particle-Level Models

The furnace residence time of pulverized coal particles mentioned in Table 2.1 or used while sizing the PC fired boiler is intimately related to particle-level processes such as drying, devolatilization, and combustion of carbon in coal (see Figure 1.3 in Chapter 1). Coal is a complex, heterogeneous mixture of organic material, moisture, and mineral matter in various compositions.

TABLE 2.1

Overall Design Guidelines for Sizing the Boiler

Parameter	Typical Value/Range	Influencing Variables
Maximum net heat input (NHI) per unit plan area of furnace	~5×10^6 kcal/hr/m²	Feed rate of fuel, GCV (Gross Calorific Value) of fuels, furnace plan area
Maximum heat liberation rate	~1×10^5 kcal/hr/m³	Fuel feed rate, GCV
Maximum burner zone heat release rate	~1×10^6 kcal/hr/m²	Fuel feed rate, GCV, burner zone volume
Maximum heat input per burner	~5×10^7 kcal/hr/burner	Fuel feed rate per burner, GCV
Maximum furnace cooling factor	~2×10^5 kcal/hr/m²	NHI/effective projected radiant heat absorbing surface
Maximum furnace exit gas temperature (FEGT) [a]	~60°C below minimum initial deformation temperature of ash	Furnace temperature, effective emissive in furnace, ash deformation temperature
Minimum furnace residence time [b]	~2 seconds	Particle size, fuel flow rate, excess air, flow patterns
Furnace bottom hopper	Designed for a hopper that is one-third full of ash/clinker	Shape and size of hopper, particle size/density of bottom ash

[a] The furnace exit plane shall be defined as the plane above the furnace nose tip or the plane beyond which the transverse tube pitching is less than 600 mm, whichever is positioned first in the flue gas path.

[b] Furnace residence time shall be defined as the residence time of the fuel particles from center line of the top-elevation coal burners to the furnace exit planes. For this purpose, the furnace exit plane shall be defined as the horizontal plane at the furnace nose tip for two pass boilers.

Pulverizing coal into smaller particles further complicates the characterization because of inherent particle size distributions and possible activation or deactivation in the process of pulverization. It is important to develop a detailed understanding of particle-level processes, such as

- Drying
- Devolatilization: volatiles, tar, decomposition of tar
- Combustion

It is also important to understand and characterize the status of the remaining ash after char combustion. These particle-level processes depend on a variety of factors, such as

- Proximate analysis: composition of volatiles, fixed char, moisture, and ash
- Ultimate analysis: elemental composition

- Particle size/size distribution
- Porosity and pore size distribution
- Ash composition/composition of minerals present
- Initial moisture content and kinetics of drying
- Kinetics of devolatilization
- Kinetics of char combustion

Some of these particle-level processes may be quantified and understood by the physical analysis of coal samples. For example, porosity and pore size distribution may be quantified using standard methods of measurement of porosity and pore size distribution (Brunauer–Emmett–Teller [BET] nitrogen and carbon dioxide sorption methods). Similarly, there are well-established methods for quantifying the elemental compositions of coal or for quantifying the minerals present in ash. However, for the remaining particle-level processes, it is essential to develop and use single-particle models for gaining a better understanding of these particle-level processes (proximate analysis, kinetics of drying, devolatilization, and combustion). Significant work on the development of such models has been published. These particle-level processes and experiments needed to gain a better understanding of these processes are discussed in Chapter 3. The mathematical models required to interpret the experimental data and obtain key kinetic characteristics of the underlying particle-level processes are also discussed in Chapter 3. These models can then be combined with engineering design/sizing models to verify some of the established practices based on experience. These models can also be combined with more detailed boiler-level models.

2.4 Boiler-Level Models

PC fired boiler-level models primarily focus on establishing a relationship between boiler configuration and operating parameters and boiler performance (fireball size and temperature, heat transferred to water walls and various internal heat exchangers, unburned carbon in ash, etc.). These relationships are essentially based on an understanding of turbulent mixing, reactive two-phase flow, exothermic heat generation, and radiative heat transfer processes taking place in PC fired boilers. The aerodynamics of the combustion zone must also be understood accurately not only to ensure that the burner jets reach the center of the furnace at the correct location, but also to stabilize the flame. The performance of a PC fired boiler is influenced by different boiler-related parameters such as furnace volume, burner type, excess air, distribution of coal/air, burner settings, and various coal-related parameters such as quality of coal, char maceral content, and particle size distribution (Hurt et al., 1998; Walsh, 1997; Chen et al., 1992). Recent advances in understanding multiphase flows, turbulence, and combustion, coupled with advances in computing power and numerical methods, provide an opportunity to develop an even better understanding of and better computational models for simulating coal fired boilers.

Computational models for simulating total boiler performance can be grouped into two broad categories, namely (1) CFD-based models and (2) reaction engineering models developed using the insights and data obtained from the CFD models. The key aspects of these two models are briefly discussed in Section 2.4.1.

2.4.1 Computational Fluid Dynamics (CFD) Models

The development of advanced industrial burners, furnaces, and boilers with the improved performance of higher efficiency and lower pollutant emissions is the major goal of combustion researchers, and furnace designers and manufacturers. To realize the goal, on one hand, the new technical concepts and novelties for different combustion routes and processes must be continuously developed. On the other hand, more efficient and economic tools, such as computer simulation using CFD technology, are also extremely important for the design and development of new advanced furnaces. In recent decades, compared with the traditionally experimental methods and physical modeling methods, computer simulation with numerical methods is considered a more effective design tool (see Ranade (2002), which discussed application of computational flow modeling for several complex industrial multiphase reactors). Over the years, CFD has evolved as a powerful design and predictive tool to simulate large utility boilers as it can handle multiple complex and simultaneous processes such as fluid flow, heat transfer, particle trajectories, and chemical reactions (Baek et al., 2014 ; Park et al., 2013; Modelinski, 2010; Choi and Kim, 2009; Belosevic et al., 2006; Pallares et al., 2005; Yin et al., 2002; Fan et al., 2001; Eaton et al., 1999; Boyd and Kent, 1986).

The CFD models represent boiler-level processes occurring in PC boilers with the help of mass, momentum, and energy conservation equations. Several approaches have been used for developing such models. These models are essentially developed to carry out detailed simulation of flow, heat transfer, and reactions occurring in PC fired boilers. These models involve the formulation and solution of strongly coupled and nonlinear partial differential equations. The solution of these models requires significant computational resources. A systematic methodology must be developed to obtain useful results from these models and then use the results for realizing performance enhancement in practice. These models are useful in evaluating the influence of hardware configurations such as location and tilt of burners, swirl at burner, locations and configurations of internal heat exchangers, air ingress, imbalance in temperatures of steam drums, heat transfer

effectiveness, etc. These models provide useful insights and facilitate performance enhancement. The development and solution of these models are
discussed in Chapter 4.

2.4.2 Reactor Network Models

As mentioned above, CFD-based models are computationally demand-
ing. The major demands on computational resources derive from the solution of momentum conservation equations. The flow in PC fired boilers
spans length scales over a large range, thus necessitating a large number of computational cells for acceptable resolution of these characteristic
length scales. Detailed CFD models are therefore not used for understanding and evaluating the performance of PC fired boilers over large space
parameters simply because of the constraints on time and other resources
required for carrying out CFD simulations. Thus, significant efforts have
been made to develop lower-order phenomenological models that will
require significantly lower computational resources than those required
for CFD models.

Such phenomenological models typically solve mass and energy conservation equations without solving momentum conservation equations.
This greatly reduces the demand on computational resources. However,
the models therefore will have to make significant approximations about
the underlying flow processes occurring in the PC fired boiler. Typically,
the development of phenomenological models is based on the understanding gained about flow processes, either from experimental measurements or
from CFD simulations. Considering the large size as well as distinctly different regions in a PC fired boiler, usually these phenomenological models
represent the PC fired boiler as a network of interconnected ideal reactors
and they are therefore called "reactor network models" (RNMs).

The number, location, size, and interconnectedness of these zones or
reactor networks are often based on CFD simulations. For each of these
zones/reactors, mass and energy conservation equations are developed
and solved. Particle-level processes can also be incorporated into these
models with the desired level of detail. For a given boiler configuration,
once such an RNM is formulated, the models can be used to evaluate the
influence of key operating parameters, such as the influence of coal composition, particle size distribution and flow rate, excess air, etc. Because
the computational demands are much lower than for the CFD models,
the RNM can be used to explore a wider parameter space. It is custom-
ary to use RNM for exploring a wider parameter space in order to shortlist the most promising operating window, which can then be evaluated
more rigorously using the CFD-based models. The details of the formulation and solution of an RNM are discussed in Chapter 5. In principle,
it is also possible to automate the formulation of an RNM from the CFD
models and even couple the RNM with the CFD model so that it can

be formulated on-line. Some comments on this possible automation and coupling with the CFD model are also included in Chapter 5.

2.5 Applying Computational Models to Practice

In addition to the development of appropriate models, solution methodology, and tools for implementing the solution methodology so as to generate simulated results, it is important to develop a systematic methodology for using those simulated results. Some of the key purposes of carrying out simulations are to

- Gain a better understanding of underlying processes.
- Establish relationships between the operating parameters and protocols with performance.
- Establish relationships between hardware design and configuration with performance.
- Evolve and evaluate ideas (operation or design domain) for performance enhancement.
- Fine-tune the most promising idea to facilitate its implementation in realizing performance enhancement.

All these points look obvious. However, it is often not very straightforward to execute these steps. Considering the complexity of PC fired boilers, computational models developed to represent such boilers invariably involve several submodels. These submodels and models involve several parameters that must be obtained from independent experiments. More often than not, in practice, the resources required for this (in terms of time, facilities, and funds) are not available to the project team. The team has to identify a smaller set of parameters that may have to be calibrated to establish a correspondence between simulated results and reality. The guidelines for model calibration are not readily available and need to be evolved based on experience and engineering judgment.

Typical performance enhancement projects involving computational modeling start with clearly identifying performance enhancement objectives. These objectives will guide what kind of models should be used. In most cases in practice, it is also essential to use multiple models to achieve key objectives. These models are used to first ensure that they capture the existing scenario with acceptable accuracy. After establishing this (if necessary, with the help of calibration), the models are then used to gain better insight into and evolve new ideas of performance enhancement, as mentioned at the beginning of this section. The overall approach of using different computational models, solving these models using digital computers, and extending the benefits of these models to realize performance enhancement in practice are discussed in Chapter 6.

2.6 Summary

Judicious use of computational models allows design and operating engineers to gain a better understanding of complex systems such as PC fired boilers. The understanding gained through this modeling usually facilitates the evolution of creative ideas for possible performance enhancement. Computational models allow numerical experiments to evaluate such new and innovative ideas as well as identify the most promising option for implementation in practice. However, the inherent complexities of PC fired boilers make the task of computational modeling these boilers quite complex. This chapter discussed the overall approach for computational modeling of PC fired boilers, which consists of multiple models, namely models for sizing/selecting configuration of PC boilers (outside the scope of this book), boiler-level models based on chemical reaction engineering (reactor network models), and computational fluid dynamics and particle-level models. Key aspects of these different models were briefly outlined. The overall approach and corresponding organization of this book are shown in Figure 2.1. The approach discussed here, combined with the details of three models as well as with the methodology of applying these models (Chapter 6), will hopefully lead to performance enhancement in practice and will further stimulate the development and application of computational modeling of PC fired boilers.

References

Aris, R. (1978). *Mathematical Modeling Techniques*, Pitman Publishing, Southport.

Baek, S.H., Park, H.Y., and Ko, S.H. (2014). The effect of the coal blending method in a coal fired boiler on carbon in ash and NOx emission, *Fuel*, 128, 62–70.

Belosevic, S., Sijercic, M., Oka, S., and Tucakovic, D. (2006). Three-dimensional modeling of utility boiler pulverized coal tangentially fired furnace, *International J. Heat and Mass Transfer*, 49, 3371–3378.

Boyd, R.K. and Kent, J.H. (1986). Three-dimensional furnace computer modeling, In *21st Symp. (Int.) Combustion*, The Combustion Institute, pp. 265–274.

Chen, J.-Y., Mann, A.P., and Kent, J.H. (1992). Computational modeling of pulverised fuel burnout in tangentially fired furnaces. In *Proc. 24th International Symp. on Combustion*, July 5-10, 1992, pp. 1381–1389.

Choi, C.R. and Kim, C.N. (2009). Numerical investigation on the flow, combustion and NOx emission characteristics in a 500 MWe tangentially fired pulverized-coal boiler, *Fuel*, 88(9), 1720–1731.

Denn, M.M. (1986). *Process Modeling,* Longman Scientific & Technical, Harlow, Essex.

Eaton, A.M., Smoot, L.D., Hill, S.C., and Eatough, C.N. (1999). Components, formulations, solutions, evaluation, and application of comprehensive combustion models, *Progr. Energy and Combustion Sci.,* 25, 387–436.

Fan, J.R., Zha, X.D., and Cen, K.F. (2001). Study on coal combustion characteristics in a w-shaped boiler furnace, *Fuel,* 80, 373–381.

Heselton, K.E. (2005). *Boiler Operator's Handbook,* Fairmont Press Inc., Lilburn, GA.

Hurt, R., Sun, J., and Lunden, M. (1998). A kinetic model of carbon burnout in pulverized coal combustion, *Combustion and Flame,* 113, 181–197.

Modelinski, N. (2010). Computational modeling of a utility boiler tangentially-fired furnace retrofitted with swirl burners, *Fuel Processing Technol.,* 91, 1601–1608.

Pallares, J., Arauzo, I., and Diez, L.I. (2005). Numerical prediction of unburned carbon levels in large pulverized coal utility boilers, *Fuel,* 84, 2364–2371.

Park, H.Y., Baek, S.H., Kim, Y.J., Kim, T.H., Kang, D.S., and Kim, D.W. (2013). Numerical and experimental investigations on the gas temperature deviation in a large scale, advanced low NOx, tangentially fired pulverized coal boiler, *Fuel,* 104, 641–646.

Polya, G. (1967). *Mathematical Discovery: On Understanding, Learning and Teaching Problem Solving,* Vol. 1, John Wiley & Sons, New York.

Ranade, V.V. (2002). *Computational Flow Modeling for Chemical Reactor Engineering,* Academic Press, London.

Speight, J.G. (2013). *Coal-Fired Power Generation Handbook,* Scrivener Publishing and John Wiley & Sons, New York.

Walsh, P.M. (1997). Analysis of carbon loss from a pulverized coal-fired boiler, *Energy Fuels,* 11, 965.

Yin, C., Caillat, S., Harion, J., Baudoin, B., and Perez, E. (2002). Investigation of the flow, combustion, heat-transfer and emissions from a 609 MW utility tangentially fired pulverized-coal, *Fuel,* 81 8), 997–1006.

3

Kinetics of Coal Devolatilization and Combustion: Thermogravimetric Analysis (TGA) and Drop-Tube Furnace (DTF)

Coal is a heterogeneous mixture of organic material, moisture, and mineral matter in various compositions. The composition of coal varies widely, depending on the time history under which the coal has undergone exposure to heat and pressure (the process is called coalification). Coals are ranked according to the coalification stage achieved and primarily are grouped into four major types in ascending order of their rank: lignite, subbituminous, bituminous, and anthracite.

The primary characterizations performed on coal—or any solid fuel—are proximate, ultimate, and ash analyses along with an estimation of the gross calorific value (GCV). The proximate analysis provides overall quantification of the volatiles, fixed char, moisture, and ash present in the coal. The ultimate analysis provides information about the carbon, hydrogen, oxygen, nitrogen, and sulfur (CHONS) present in the coal. The ash analysis provides details about ash constituents such as the oxides of silicon, aluminum, calcium, iron, magnesium, potassium, sulfur, etc. The GCV provides the total heat content of the coal along with the heat of condensation of water in the combustion products. The GCV is primarily estimated through experiments; otherwise, various empirical correlations such as Dulong's formula and its variant can be used (Sarkar, 2009; Parikh et al., 2005). The net calorific value (NCV), which is the net heat content of the coal, is calculated from the GCV based on the moisture and sometimes hydrogen present in the coal (American Society for Testing of Materials [ASTM] Standard D5865).

The organic matter present in coal is heterogeneous in composition, consisting of different types of macerals (microscopic components). During the combustion of coal particles, each type of maceral exhibits a different behavior in terms of swelling, volatiles yield, char structure, and reactivity. Therefore, more recently coals have also been classified according to the nature of their microscopic maceral/petrographic analysis. The organic material is divided into three groups (Speight 2013): vitrinite (woody materials), exitnite (spores, resins, and cuticles), and inertinite (oxidized plant material). For most coals, the vitrinite group is the major constituent. Microscopic studies show that coal may be made up of a single maceral or a combination of macerals (Gupta et al., 1999). Due to the optical properties of vitrinite, the vitrinite reflectance is used as an indicator of rank. Figure 3.1 shows typical characteristics of various coals. Classification of coal based on vitrinite reflectance indicates that peat has the lowest vitrinite reflectance (R_{max} = 0.2%), and anthracite has the highest vitrinite reflectance (R_{max} = 7%) and thus the highest rank. The other key characteristics, such as moisture content, carbon content, volatiles content, and gross calorific value, are also indicated in Figure 3.1.

There are coking and non-coking types of coal. Steel industries are the primary consumers of coking coal, and power utilities as well as other process industries typically use non-coking coal. The non-coking coals are classified based on their calorific value. Normally in India D, E, and F coal grades are available to utility industries; the E grade, which has an ash content greater than 35%, is commonly fired in most PC boilers operating in India. Typical characteristics of the Indian non-coking coals are shown in Figures 3.2, 3.3, and 3.4.

A useful graphical representation of various types of solid fuels can be obtained from a Van Krevelen diagram. This is essentially based on a relationship between the ratio of atomic hydrogen to carbon (H/C: hydrogen index) and the ratio of atomic oxygen to carbon (O/C: oxygen index) present in coal on a dry and ash-free basis (daf basis). Typically, anthracite has the lowest oxygen content (max ~0.2), which progressively increases in lignite

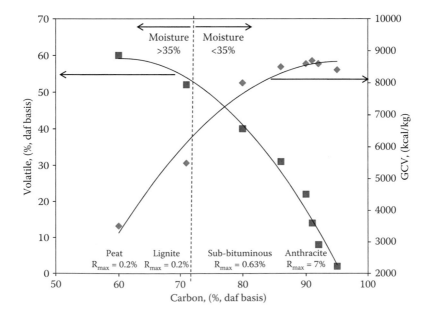

FIGURE 3.1
Typical characteristics of different coals (data from Diessel, 1992). (daf: dry ash-free basis, R_{max}: maximum vitrinite reflectance.)

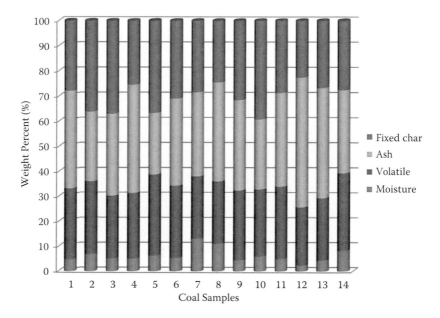

FIGURE 3.2 (See Color Insert.)
Typical characteristics of Indian D (Sample no. 2-5, 7-10, 14) & E (Sample no. 1, 6, 11–13) type coal (data from Mishra, 2009).

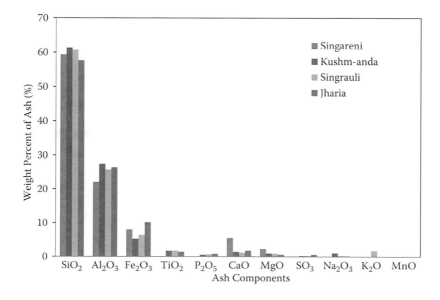

FIGURE 3.3 (See Color Insert.)
Typical ash composition of Indian coal (data from Chandra, 2009).

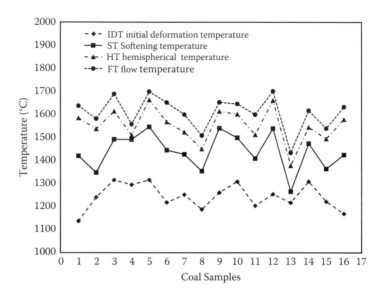

FIGURE 3.4
Ash fusion temperature data of non-coking coal (D, E, and F grade) commonly found in Orissa, India (data from Kumar and Patel, 2008).

(max ~0.4) and peat (max ~0.5). An approximate relationship between these ratios (on a daf basis) can be represented as

$$10(H/C) = 1.2 - 1.4 \ (O/C) + 0.7 \tag{3.1}$$

Biomass has a higher oxygen content, and the above equation may hold approximately for biomass as well.

Simple analyses such as proximate, ultimate, and ash analyses discussed above do not provide any information on the rates of processes such as drying, devolatilization, and char combustion of coal particles (see Figure 1.3 in Chapter 1). These processes are quite complex and depend on many factors. The loss of moisture takes place very rapidly in a typical PC fired boiler. The kinetics of drying therefore are not discussed here because they are not relevant for the modeling of coal combustion in a PC fired boiler. As discussed in a previous chapter (Chapter 2), it is important to quantify devolatilization and combustion kinetics in order to develop a computational model for a PC fired boiler. Coals with similar properties such as GCV and ash content may show very different devolatilization and combustion behaviors. The thermogravimetric analyzer (TGA) and drop-tube furnace (DTF) are commonly used to understand and quantify the rates of devolatilization and char combustion. Particle-level models are used to process TGA and DTF data. The details of such characterizations using TGA and DTF are discussed in this chapter. Before that, a general discussion of these processes is included in Section 3.1.

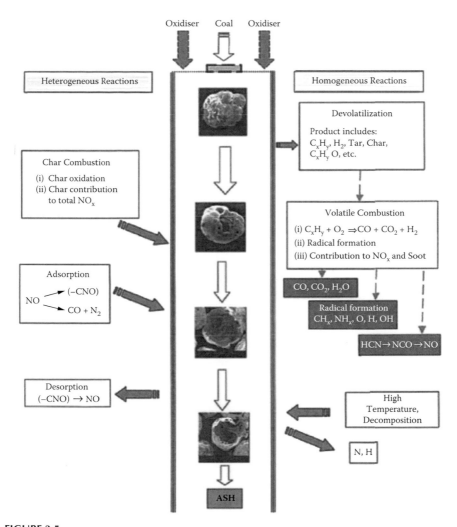

FIGURE 3.5

Schematic of coal combustion processes. Reprinted from *Proceedings of Combustion*, 28(2), Williams, A., Pourkashanian, M. and Jones, J.M. 'The Combustion of coal and some other fuels', 2141–2162(2000) with permission from Elsevier.

3.1 Coal Devolatilization and Combustion

The process of the combustion of pulverized coal particles starts with the heating of coal particles, followed by loss of moisture (100°C to 150°C), release of volatile materials (280°C to 350°C), and combustion of char (fixed carbon). Ash is left behind as a final solid product, with some unburned carbon trapped in the ash matrix. The general processes that take place during coal combustion are schematically shown in Figure 3.5.

Devolatilization is the thermal decomposition of coal where the volatile material present in the coal is released in gaseous form, such as CO, CH_4, NH_3, H_2S, tar, etc. Depending on the temperature and heating rate experienced by the coal particles, the devolatilization products and their yields can vary. Commonly, TGAs are used to quantify the rates of devolatilization. It should be noted that a typical TGA allows for rather low heating rates (~<1K/s) of coal particles. In real PC fired boilers, particles may experience orders of magnitude higher heating rates, which may in turn influence the devolatilization process. Occasionally, fast pyrolyzers, which may allow high heating rates (up to 10^4K/s), are used for quantifying devolatilization processes. Another widely used experimental set-up for this purpose is the drop-tube furnace (DTF). The scope of this chapter is restricted to discussions of TGA and DTF. Both of these instruments may be attached to mass spectroscopy (MS) or gas chromatography (GC) to analyze the reaction products emanating from the devolatilization and combustion processes. An estimation of the kinetics of devolatilization and char combustion using TGA and DTF is discussed in Sections 3.2 and 3.3, respectively. Some general comments and a brief review of devolatilization and char oxidation (combustion) models are included here.

There are a number of approaches for modeling the complex devolatilization process (summarized in Table 3.1). The single reaction model (SRM) and multireaction model (MRM) are two broad categories under which these models can be categorized. A single first-order reaction is the most commonly used SRM, while a distributed activation energy model (DAEM) is the most widely used MRM. The more exhaustive and advanced methods based on a functional group approach, such as functional group depolymerization, vaporization, cross-linking (FG-DVC, Solomon et al., 1988), FLASHCHAIN (Niksa and Kerstein, 1991), and chemical percolation devolatilization (CPD, Fletcher and Kerstein, 1992), are quite useful in predicting the volatile yield and its composition at various heating rates and operating pressures. The details of each model can be found in the literature cited in Table 3.1. Such models are supported based on numerous coal structure-level information and measurements. There are advancements on such models that can interpolate devolatilization behavior for any coal based on the existing library of coal. In the absence of more detailed information, a simple SRM model can be used.

The process of char combustion is of central importance in industrial PC fired boiler applications. The overall process of coal transformation is quite complex and involves many distinct steps. There are many parameters, such as particle size, char mineral content, and fragmentation of the char particle, that affect the reaction rate (see Figure 1.3 in Chapter 1). The rate-limiting step in the combustion of a char particle may be either a chemical reaction, gaseous diffusion to the particle, or some combination of these two factors. Generally, three zones are defined, namely

TABLE 3.1

Kinetic Expressions Used for Devolatilization

Mechanism	Kinetic Expression	Source of Information and Remarks
Single step $Coal \xrightarrow{k_1} X(volatile) + (1-X)(Char)$	$k_1 = A_1 \exp(-E_1/RT)$ $dV/dt = k(v^* - v)$	Howard et al. (1987) Devolatilization rate may depend upon the specific coal and reaction conditions
Two Step $Coal \xrightarrow{k_1} Volatile\ 1 + Residue\ 1$ $Coal \xrightarrow{k_2} Volatile\ 2 + Residue\ 2$	$\dfrac{dx_{coal}}{dt} = -k_1 x_{coal} - k_2 x_{coal}$ $k_1 = A_1 \exp(-E_1/RT)$ $k_2 = A_2 \exp(-E_2/RT)$	Kobayashi et al. (1976) $A_1 = 3.7 \times 10^5$ (1/s) & $E_1/R = 8{,}857$ (K) $A_2 = 1.46 \times 10^{13}$ (1/s) & $E_1/R = 30{,}200$ (K) Generality limitation: as one step
Distributed activated energy model (DAEM) Number of independent parallel, first-order reactions	$dv_i/dt = k_{o,i} \exp\left(\dfrac{-E_i \pm \sigma_{E,i}}{RT_p}\right)(v_o - v_i)$	Miura (1995); Serio et al. (1987)
Functional Group (FG)	Based on elementary reaction	Gavalas (1982)
FG – Depolymerization – vaporization – cross-linking model (FG – DVC)	Network pyrolysis computer code	Solomon and Fletcher (1994)
FLASHCHAIN	Network pyrolysis computer code	Niksa and Kerstein (1991)
Chemical percolation devolatilization model (CPD)	Network pyrolysis computer code	Juntgen and Van Heek (1979); Smith et al. (1993)

Source: Adapted from Proceedings of the Combustion Institute, 28(2), Williams, A., Pourkashanian, M., and Jones, J.M. *The Combustion of Coal and Some Other Solid Fuels*, pp. 2141–2162 (2000). With permission from Elsevier.

Zone I: Chemical reaction is the controlling step. It occurs at low temperatures or with small particles. As the chemistry is slow, the gas species are transported fully inside the coal particle. The char combustion rate shows a reaction order of 0.6 to 1.0 toward oxygen. The typical temperature is <600°C

Zone II: Both chemical reaction and pore diffusion control. Gas species are partially transported inside the coal particle. The char combustion rate shows a reaction order of 0.6 to 1.0 toward oxygen. The typical operating temperature is 600°C to 1,400°C.

Zone III: Bulk mass transfer of oxygen controls, or the particles are of large size. As the chemistry is fast, the reaction takes place close to the particle external surface area. The char combustion rate shows a reaction order close to 1.0 toward oxygen. The typical operating temperature is >1,400°C.

It has been common practice to relate experimental char burning rates to the external char surface area, even when pore reactions can occur. The resulting rate of reaction is called "global" because it incorporates the influence of the pore surface area.

The commonly used methods for characterizing char oxidation are TGA and DTF or its variants such as entrained flow reactor, laminar flow reactor, or wire mesh reactors. A brief summary of these methods is included in Table 3.2. A review by Solomon et al. (1992) may be referred to for details on experimental methods to study coal pyrolysis/devolatilization.

The TGA- and DTF-based experiments provide useful information on the yield of volatile products and rates of devolatilization and char combustion (reactivity of char). Different basic char burnout models are summarized in Table 3.3 (from Williams et al., 2000). Numerous enhancements to these basic models have been proposed (mainly to capture reduced oxidation rates in the high conversion range) and are summarized below:

- Based on petrography, ash content, and heterogeneity of material
- Thermal annealing
- Extinction and near-extinction phenomena

Hurt and co-workers (2003, 1998) developed a statistical model that addresses the heterogeneity in both reactivity and density. Thermal annealing of the fuel matrix has also been observed (Hurt and Gibbins, 1995), which may affect char reactivity. However, this effect is thought to affect primarily the initial stages where the maximum temperatures are reached, rather than affecting the final stages of the particle's combustion history (Sun and Hurt, 2000). The

TABLE 3.2

Brief Summary of Experimental Techniques for Coal Characterization

	Experimental Technique: Thermogravimetric Analysis (TGA)	
Refs.	**Operating Condtions**	**Key Features**
Sarkar et al. (2014)	Non-isothermal TGA: Sample mass = 20 mg, heating rate (10K/min) and particle size (~75 μm), air flow rate = 50 ml/min, temperature = 1,023K. DTF: φ = 100 mm, L = 2,500 mm, and T = 1,273K, 20% excess air.	Co-combustion studies of sawdust, sawdust char, and coal in TGA and drop-tube furnace.
Farrow et al. (2013)	Pyrolysis heating rate (50 K/min) and particle size (125–250 μm (biomass); 53–75 (coal)), air flow rate = 50 ml/min, temperature = 1,123K. Oxidation studies were performed as isothermal temperature. DTF: T = 1,173–1,573K, N₂/CO₂/oxy-fuel.	Influence of co-firing of biomass and coal under oxy-fuel conditions, sawdust, pinewood, and a South African coal in TGA and DTF.
Yuzbasi and Selcuk (2011)	Non-isothermal TGA: Sample mass = 12 mg, heating rate (40K/min) and particle size (<100 μm), N₂ and CO₂ = 70 ml/min (pyrolysis), air flow rate = 45 ml/min (oxidation), mixed in various proportion for oxy-fuel combustion studies, temperature = 1,223K.	Lignite and olive residue and their 50-50% blend under air and oxy-fuel conditions.
Seo et al. (2011)	Non-isothermal TGA: Sample mass = 800 mg, heating rate (5–30K/min) and particle size (~60–70 μm), air flow rate = 3.0 l/min, temperature = 1,173K.	TGA experiments on coal with online gas analyzer.
Sahu et al. (2010)	Non-isothermal TGA: Sample mass = 20 mg, heating rate (10K/min) and particle size (~75 μm), air flow rate = 50 ml/min, temperature = 1,023K.	Co-combustion of coal and biomass to estimate kinetic parameters and ignition index.
Jones et al. (2005)	Non-isothermal TGA: Sample mass = 15 mg, heating rate (25K/min) and particle size (~75–90 μm), air flow rate = 50 ml/min, temperature = 1,173K.	Devolatilization characterstics of 3 ranks of coal and pinewood blends (25%–75%).
Cloke et al. (2002)	Non-isothermal TGA	Intrinsic reactivity of 14 coal samples.
Alonso et al. (2001)	Non-isothermal, HR = 25K/min; sample mass 13 mg; air 50 ml/min (for oxidation, 1,273K); N₂ 50 ml/min (for oxidation, 1,173K).	Effect of vitrinite and inertinite content on coal combustion.
Morgan et al. (1986)	Standardization of sample mass (~5 mg), heating rate (20K/min) and particle size (< 75 μm).	Predicted effect of rank on char reactivity.
Cumming (1984)	Non-isothermal, HR = 15K/min; sample mass-20 mg; air-75 ml/min.	Proposed reactivity assessment via a weighted mean activation energy. Tested 22 coal samples of all ranks.

(Continued)

TABLE 3.2 (*Continued*)

Brief Summary of Experimental Techniques for Coal Characterization

Experimental Techniques: Drop-Tube Furnace (DTF)/Entrained Flow Reactor (EFR)/ Laminar Flow Reactor (LFR)		
Refs.	**Operating Condtions**	**Key Features**
Wang et al. (2014a)	$\phi = 50$ mm, L = 1,300 mm; T = 1,373K; solids fuel rate = 25 g/hr; Primary + Secondary air flow rate = 22 l/min.	Experiments on RDF, rice husk, straw, and coal.
Wang et al. (2014b)	$\phi = 50$ mm (outer), $\phi = 38$ mm (inner), $L = 1,300$ mm; T = 1,373K (max temperature),	Bituminous coal under three oxy-fuel condtions (under O_2-CO_2 gas environment) at 1,100°C. Measured NOx formation under various conditons.
Li et al. (2013)	$\phi = 50$ mm, $L = 1,300$ mm; T = 1,073K; 1,273K; 1,473K; 1,673K; N_2 as inert gas flow rate = 500 ml/min.	Co-pyrolysis behavior of sawdust and Shenfu bituminous coal blends in various ratios were tested.
Haykiri-Acma et al. (2013)	$\phi = 50$ mm, $L = 720$ mm; T = 873K; 973K; 1,073K; 1,173K.	Low tempeature (600–900°C) experiments on biomass oxidation.
Chi et al. (2010)	$\phi = 500$ mm (outer), $\phi = 200$ mm (inner), $L = 1,500$ mm (ignition section), L = 2,000 mm (burnout section), total length = 3,500 mm; T = 1,123K, and 1,623K (max temperature); Fuel feeding rate = 1.2 kg/hr.	Ignition behaviors of pulverized coals and coal blends in a drop-tube furnace using a flame monitoring system.
Chen and Wu (2009)	T = 1,400K, Carrier gas (N_2) = 2 L/min, Reactor gas (N_2) = 2 L/min.	Pulverized coal and rice husk blends.
Jimenez et al. (2008)	$\phi = 78$ mm, $L = 1,600$ mm; T = 1,073K; 1,203K; 1,313K; 1,448K.	Devolatilization and oxidation of pulverized biomass (*Cynara cardunculus*) to estimate kinetic parameters from DTF data.
Yoshizava et al. (2006)	$\phi = 42$ mm, $L = 800$ mm; T = 400–1523K; HR = 10^2–10^3K/s.	Swelling characteristics of 11 types of coal.
Ballester and Jimenez (2005)	$\phi = 78$ mm, $L = 1,600$ mm; T = 1,313K; 1,448K; 1,573K; 1,723K.	Anthracite coal at various operation temperatures and O_2 concentrations. Proposed method to estimate the kinetic parameters from drop-tube furnace data.

TABLE 3.2 (*Continued*)

Brief Summary of Experimental Techniques for Coal Characterization

Experimental Techniques: Drop-Tube Furnace (DTF)/Entrained Flow Reactor (EFR)/ Laminar Flow Reactor (LFR)		
Refs.	**Operating Condtions**	**Key Features**
Zhang et al. (2005)	ϕ = 200 mm, L = 2,500 mm; T = 1,523K.	Used AUSM turbulence chemistry model of char combustion to predict influence of particle temperature fluctuation on char combustion rate.
Cloke et al. (2003)	T = 1,573K; ϕp = 600 ms; 5% O_2.	Effect of inclusion of char morphological data in CBK model.
Hurt et al. (1998)	T = 1,423K; τp = 500 ms.	Char conversion data simulated with models based on CBK and global rate.
Card and Jones (1995)	ϕ = 25 mm, L = 2,500 mm; T = 1,573K; τp = 2.5 s.	Used light-scattering technique to study coal combustion and fly ash formation in DTF.
Nandi and Vleeskens (1986)	ϕ = 25 mm, L = 1,260 mm; T = 1,573–1,773K; τp = 1.2 s.	Influence of ash, vitrinite, and inertinite on char burnout in DTF.
Hindmarsh et al. (1995)	Wire mesh reactor and entrained flow reactor.	Comparison of pyrolysis carried out in wire mesh reactor and entrained flow reactors.

Note: Φ and L are internal diameter and length of DTF tube, respectively.

presence of inorganic material may also reduce oxidation rates through different mechanisms (Zolin et al., 2001). As char burns, this factor will have increasing influence as the ash content of the particles increases with burnout and particle residence time. Extinction and near-extinction phenomena may cause an abrupt decrease in the oxidation rate of individual particles (Zolin et al., 2001; Hurt et al., 1998; Essenhigh et al., 1999), thus contributing to a reduction in the global conversion rate. The details of such models can be found in Hurt et al. (1995, 1998, 2000, 2003).

Smith (1982) has proposed generalized expression for the intrinsic reactivity of char for various types of coal as

$$\text{Rate}(\text{gm/cm}^2/\text{s}) = 305\, e^{\left(-\frac{42.8}{RT_P}\right)} \tag{3.2}$$

where T_p is the temperature of the particle in Kelvin and R is the universal gas constant.

TABLE 3.3

Char Burnout Models

Source and Key Features	Data
Baum and Street (1971)	E and A vary from coal to coal

$$\frac{dm_c}{dt} = -\pi D^2 \rho RT \frac{X_{O_2}}{M_{O_2}} \left(R_{diff}^{-1} + R_c^{-1} \right)^{-1}$$

where m_c is the mass of the coal particle, D is diameter of the particle, ρ is density of particle, R is the universal gas constant, T is the temperature, X is the mole fraction of species, M is the molecular weight of species, R_c is the chemical rate coefficient/unit external surface area, R_{diff} is the diffusional reaction coefficient.

In these expressions, the rate of mass loss by combustion depends on particle density, diameter, and the ratio of reacting surface to external surface area of the particle.

Intrinsic reaction rates: Smith (1982)	$E = 179.4$ kJ/mol
	161 ± 6 kJ/mol

$$\rho_i = \rho_o / \eta \gamma \sigma_a Ag[C_g(1-X)]^m$$

where ρ_i is the intrinsic reactivity, ρ_o is the observed reactivity, η is the effectiveness factor, γ is the characteristic size, σ_a is the apparent density, A_g is the total surface area, C_g is the oxidizer concentration, m is the true reaction order, and X is the fractional burnout.

Hampartsoumian et al. (1998)	$E = 167.3$ kJ/mol

$$R_c = \exp^{(-89)} \sigma^{(-7.5)} A_g^{(-0.5)} C^{(3.5)} T_p^{(9.5)}$$

$$[1.4(vit_m + 0.83vit_{ps})] - [0.6(ln_R + 1.6ln_{LR})]$$

where vit_m and vit_{ps} are the fractions of matrix and pseudo vitrinite, respectively;

ln_R and ln_{LR} are the fractions of low reflectance and high reflectance inertinite, respectively; and C is carbon content (%).

Hurt et al. (1998)	$n = 0.5$
	$E = 146.3$ kJ/mol

$$R = \eta A_o S\, e^{-E/RT_p} P_o^n C_P$$

where A_o is the preexponential factor, S is the total surface area/mass, P_o is the oxygen partial pressure, n is the apparent reaction order, and C is the carbon mass in the particle.

Extended Resistance Equation: Essenhigh and Mescher (1996)

$$\frac{1}{R_s} = \frac{1}{k_{DYOX}} + \frac{1}{\in k_{aYOX}} + \frac{1}{\in k_d}$$

where R_s is the specific reaction rate at the particle surface, k_D is the oxygen diffusion velocity constant, k_a and k_d are the adsorption and desorption velocity constants, respectively, and $\in$ is the reaction penetration factor.

Source: Reprinted from *Proceedings of the Combustion Institute*, 28(2), Williams, A., Pourkashanian, M., and Jones, J.M., pp. 2141–2162 (2000). With permission from Elsevier.

Typically, char oxidation studies are performed on a drop-tube furnace because it resembles the operating conditions of the PC fired boiler, such as high heating rate and short residence time (excluding turbulence). Many times, due to the simplicity of the TGA instrument, it is also used for char oxidation studies. Details of coal characterization using TGA and DTF are discussed in Sections 3.2 and 3.3, respectively.

3.2 Coal Characterization Using Thermogravimetric Analysis (TGA)

The thermogravimetric analyzer (TGA) allows a user to monitor the change in weight of a known quantity of solid sample as a function of time and temperature. The TGA is routinely used to characterize coal and a variety of other materials, including biomass, medical wastes, waste car tires, printed circuit boards, and sewage sludge (Gašparoviè et al., 2012). Typically in a TGA experiment, a small coal particle is placed on a load cell in an enclosure with a controlled atmosphere. The weight of the particle is continuously monitored while the particle is either heated at a specified rate or the temperature is maintained for a desired period of time. Devolatilization is characterized by maintaining an inert atmosphere (e.g., N_2), while char combustion is characterized in the presence of dilute oxygen (6% to 7%), air, or pure oxygen. Typically, only a small amount (2 to 20 mg) of sample is required, and the TGA is comparatively simple to operate. There are two methods used to perform TGA studies:

1. Isothermal, where the sample is maintained at a constant temperature
2. Non-isothermal, where the sample is heated at a constant rate (~10 to 60K/min)

Standards for carrying out proximate analyses for coals using TGA from both the American Society for Testing and Materials and the International Organization for Standardization (ASTM D7582–12 and ISO 17246:2010) have been prescribed. These standards provide proven methods to determine the components of the coal sample and also to ensure the reproducibility of results.

Isothermal methods are typically significantly more time consuming and are usually used to characterize up to the first 50% of weight loss. For example, at 500°C and with air as the medium, typical residual unburned char may take 3 to 4 hr to reach 50% conversion and a further 8 to 10 hr to reach 99% conversion. Non-isothermal methods have the advantage of being able to achieve complete conversion in a single and shorter experimental time. This also helps in comparing various types of fuel and their reactivities in a short period of time. The processing of data obtained from non-isothermal experiments is more complex than that obtained from isothermal experiments. The devolatilization and char combustion kinetics are usually estimated from the observed weight loss behavior during the process of heating the particle at a constant rate (usually with a heating rate of 5 to 20 K/min) from 473K to 1,173K. Virtually all chars of interest for pulverized fuel combustion are converted by 1,173K, which is within approximately 1 to 3 hr. Thermogravimetric (TG) and differential TG (DTG) curves are generally analyzed to estimate characteristic temperatures such as peak temperature at the maximum weight loss rate, temperature at which 50% burn-off or devolatilization occurs, the burnout temperature where the DTG profile reaches a 1% combustion rate at tail end of the profile, maximum dw/dt (%/min), and kinetics of combustion and devolatilization.

3.2.1 Typical Example of a TGA Experiment

A typical TGA experiment and the data obtained from it are illustrated here. A sample of a typical subbituminous type high-ash coal commonly used in thermal power plants in India is used for this experiment. The proximate analysis of the coal was (VM: volatile matter; FC: fixed carbon):

moisture:ash:VM:FC :: 12:41:23:24

and the ultimate analysis of the coal was

C:H:N:S:O :: 37:2.26:0.85:0.33:6.53

The coal sample (~4.8 to 5 mg) in the size range between 70 and 90 µm was selected in order to avoid the influence of mass and heat transfer during the process of thermal decomposition of the sample (Aboyade et al., 2012). The sample was placed in a platinum crucible of 100-µL capacity, suspended from the balance unit on a hang-down hook. The air-cooled furnace with a maximum temperature of 1,200°C was heated with four infrared (IR) bulbs. Previous studies have indicated that parameters such as heating rate (5 to 50 K/min), mass of sample (<20 mg), flow rate of inert/air to the sample, etc. do not influence the overall trends of the results (Russell et al., 1998). Following the suggestion of Morgan et al. (1986) to avoid ignition of the sample, we used a coal sample of 5 mg that was heated with a 20 K/min heating rate in this illustrative example. A temperature-ramped non-isothermal method as discussed in Russell et al. (1998) was used to characterize the coal.

For characterizing devolatilization, nitrogen (N_2) gas was passed through the furnace section over the sample. The N_2 flow rate was kept at 20 ml/min for the sample and 40 ml/min for the balance. The sample was heated with a ramp of 20 K/min from a temperature of 30°C to 110°C. It was then kept at 110°C for 10 min so that the moisture present in the coal evaporates.

The sample was further heated to 900°C with a ramp of 20K/min so that the volatile material is released from the coal, leaving behind the char and ash. For characterizing char oxidation, the furnace was quickly cooled to 400°C in the presence of N_2, and then the air flow was purged over the sample at a flow rate of 100 ml/min and 20 ml/min N_2 for the balance. The sample was then heated to 900°C with ramp of 20K/min so that the char is oxidized. Data analysis was achieved using thermal analyzer software (Thermal Analyst 5000, TA Instruments, New Castle, Delaware). Typical TGA data after post-processing are shown in Figure 3.6. Processing this data to estimate devolatilization and char combustion kinetics is discussed in Section 3.2.2.

3.2.2 Processing TGA Data

TGA data have been used to extract the kinetics of devolatilization since the beginning of the twentieth century (Caballero and Conesa, 2005). Then, the approach mostly consisted of fitting the TGA data via linear regression, wherein the differential or integral form of the rate equation was manipulated to obtain a straight-line plot from which the unknown parameters were extracted. As a first approximation, a single first-order step (reaction)

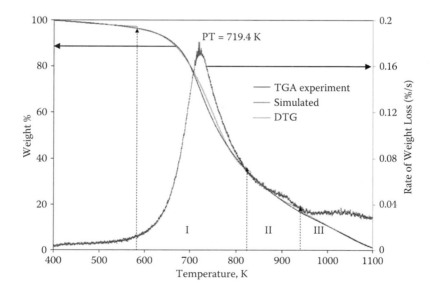

FIGURE 3.6 (See Color Insert.)
Typical results for devolatilization of high-ash coal (TGA: weight loss, DTG: rate of weight loss).
Moisture : ash : VM : FC :: 12 : 41 : 23 : 24
C : H : N : S : O :: 37 : 2.26 : 0.85 : 0.33 : 6.53

model (one pair of Arrhenius parameters: frequency factor and activation energy) is fitted to represent the TGA data. Iso-conversional methods (methods wherein the kinetic parameters are derived without assuming any reaction mechanism) such as the Kissinger–Akahira–Sunose Model, the Ozawa–Flynn–Wall model, and the Friedmann method can also be used to derive the kinetic parameters (Damartzis et al., 2011). If the process cannot be described adequately by a single first-order reaction model, either n^{th}-order single reaction models or multistep (multireaction) models (MRMs) are used. The most widely used MRMs include the multiple pseudo-component first-order reactions and the DAEMs.

The basic idea of the DAEM is that there are an infinite number of reactions occurring simultaneously during the devolatilization/pyrolysis of coal. This distribution of activation energies corresponding to these infinite reactions is often represented by a distribution function of activation of energy $f(E)$. This is a common feature in all DAEMs. Several variants of DAEM are used in which $f(E)$ and the reaction orders differ. Here, a generalized equation has been formulated from which variations of various SRMs and MRMs to simulate the TGA data can be deduced. A typical reaction occurring in a TGA experiment may be represented as a first-order reaction. Kinetics of devolatilization of the ith reaction are represented as

$$\frac{dV_i}{dt} = k_{vi}(V_i^* - V_i) \tag{3.3}$$

where V_i is the mass of volatiles released at time t, V_i^* is the total mass of volatiles originally available for the ith reaction, k_{vi} is the kinetic rate constant; the effect of temperature on the rate constant is assumed to follow the Arrhenius law:

$$k_{vi} = A_v e^{\left(-\frac{E_{vi}}{RT}\right)} \tag{3.4}$$

The simplest model to process TGA data assumes just one single reaction and estimates the frequency factor and activation energy using Equations (3.3) and (3.4). It is also possible to model it using multiple pseudo-components, each proceeding with a single activation energy (M-SRM or M-n-SRM) contributing (with a weightage w_i) to the devolatilization. The other extreme is to represent the devolatilization by infinite parallel reactions having activation energies ranging from zero to infinity. In this approach (the distributed activation energy approach), the amount of volatiles (dV) that would release having an activation energy between E_v and ($E_v + dE_v$) can be written as

$$dV = V^* \int_{E_v}^{E_v + dE_v} f(E_v)d(E_v) \tag{3.5}$$

A generalized equation representing devolatilization can be formulated for a non-isothermal TGA experiment conducted at heating rate β (K/min),

$$1-\frac{V}{V^*} = \sum_{j=1}^{m} w_j \int_0^{\infty} \left(e^{\int_{T0}^{T} \frac{Av}{\beta} e^{\left(-\frac{Ev}{RT}\right)} dT} \right) f_j(E_v) dE_v \quad \text{for} \quad n=1 \quad (3.6)$$

$$1-\frac{V}{V^*} = \sum_{j=1}^{m} w_j \int_0^{\infty} \left(1-(1-n)e^{\int_{T0}^{T} \frac{Av}{\beta} e^{\left(-\frac{Ev}{RT}\right)} dT} \right)^{\left(\frac{1}{1-n}\right)} f_j(E_v) dE_v \quad \text{for} \quad n \neq 1 \quad (3.7)$$

where j is the number of reactions, m is the total number of reactions, n is the order of the reaction, and w is the weighing factor.

For solving Equations (3.6) and (3.7), the distribution function of the activation energy $f(E)$ must be assumed. Various distribution functions have been used; some of the most commonly used ones are the Dirac delta distribution (for SRMs), logistic distribution, and Gaussian distribution. The Gaussian and logistic distributions with standard deviation σ and mean activation energy E_o can be written as Equations (3.8) and (3.9), respectively:

$$f(E) = \frac{1}{\sigma\sqrt{2\pi}} e^{\left(-\frac{(E-E_0)^2}{2\sigma^2}\right)} \quad (3.8)$$

$$f(E) = \frac{\pi}{\sigma\sqrt{3}} \frac{e^{\left(-\frac{\pi}{\sqrt{3}}\left(\frac{E-E_0}{\sigma}\right)\right)}}{\left(1+e^{\left(-\frac{\pi}{\sqrt{3}}\left(\frac{E-E_0}{\sigma}\right)\right)}\right)^2} \quad (3.9)$$

These distributions are shown in Figure 3.7 for a specific set of parameters to illustrate their differences.

Using such a generalized formulation, key parameters of devolatilization occurring in TGA experiments can be obtained by carrying out optimization to minimize the following objective function:

$$F = \sum_{z=1}^{N_z} \left(y_{exp_z} - y_{sim_z} \right)^2 \quad (3.10)$$

The quality of fit is gauged by calculating the mean average error according to Equation (3.11). The standard deviation of the mean average error was also calculated to compare the quality of fit.

$$\text{Mean average error (\%)} = \sum \sum_{z=1}^{N_z} \frac{(y_{exp_z} - y_{sim_z})}{y_{exp_z} N_z} * 100 \quad (3.11)$$

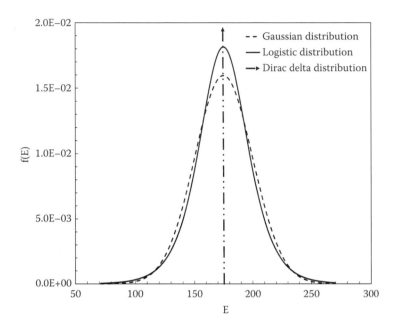

FIGURE 3.7
Comparison of Dirac delta, logistic, and Gaussian distribution ($E0 = 175$ and $\sigma = 10$ for the latter two distributions).

Typical application of DAEM to the TGA data obtained from various sources is shown in Figure 3.8 (taken from Jain et al., 2014).

It should be noted that inherent assumptions in this analysis are that a change in particle temperature is assumed to be proportional to the heating rate, and that gas-phase reactions are neglected.

The methodology to solve the DAEM equations can be categorized into two groups: the first one is a methodology wherein the functional form of the distribution of activation energy, $f(E_v)$, is assumed, and the second is a method proposed by Miura (1995) and Miura and Maki (1998) in which the functional form of the distribution of activation energy need not be assumed. Cai et al. (2011) have recently carried out a critical study of the Miura–Maki method and concluded that the method may lead to errors in the estimation of ko. A study by Fiori et al. (2012) pointed out that the Miura–Maki method could be applied to a restricted range of conversion. Some of the key efforts to process TGA data using the DAEM approach can be found in Jain et al. (2014).

Despite sophisticated models such as DAEM, more often than not, TGA data are still processed with SRMs. One of the reasons is that SRMs are relatively straightforward to implement in the more complex models of simulating PC fired boilers (see Chapter 4). In order to improve the fit with the experimental data, it is often necessary to divide the overall range of temperature used in a TGA experiment into different parts and then fit different parameters for each sub-range of temperatures. A typical TGA is shown in

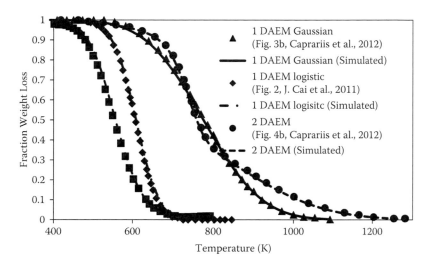

Legend:
- ▲ 1 DAEM Gaussian (Fig. 3b, Caprariis et al., 2012)
- — 1 DAEM Gaussian (Simulated)
- ◆ 1 DAEM logistic (Fig. 2, J. Cai et al., 2011)
- –·– 1 DAEM logisitc (Simulated)
- ● 2 DAEM (Fig. 4b, Caprariis et al., 2012)
- – – 2 DAEM (Simulated)

FIGURE 3.8
Comparison of published TGA data with the simulated results (Jain et al., 2014).

Figure 3.6, which shows three curves: (i) TG curve plotted as the change in the normalized weight of volatile (%) with respect to temperature and super-imposed with (ii) DTG curve (%/sec), and (iii) simulated TGA results. It can be seen from this figure that the devolatilization starts around 600K, which was commonly observed for other types of coals. The rate of weight loss curve (DTG: differential thermogravimetric curve) also confirms the same. The DTG peak temperature (PT), which represents the temperature at the maximum rate of weight loss of 0.1812%/s, was found to be 719.4K. The temperature at which the 50% weight loss occurs was recorded as 767.25K.

The TGA model was simulated and fitted to experimental data to obtain kinetic parameters for devolatilization. The model could not capture the complete section of weight loss profile for devolatilization with a single set of A_v and E_v. The overall range was therefore split into three sub-ranges (indicated by Roman numerals in Figure 3.6). The first sub-range was 600K to 830K, in which the devolatilization starts and achieves more than 60% weight loss. The second sub-range of 830K to 940K accounts for ~20% weight loss. The third sub-range was 940K to 1,100K, where the remaining 20% weight loss takes place. The kinetic parameters estimated for these three sub-ranges are listed in Table 3.4. It can be seen from Figure 3.6 that the fit achieved using the three sub-ranges is quite adequate. For illustration purposes, the splitting of the overall range into different sections was done manually here. However, optimization tools can be used to carry out this operation.

For char combustion, further assumptions are needed. Some of these are (1) the char combustion reaction is based on the external surface area of the coal particle; (2) char oxidation is first order, with oxygen partial pressure at the surface (which may be taken to be the same as that of the bulk gas); and (3) the size of

TABLE 3.4

Parameters Obtained from TGA Data

Devolatilization Kinetic Parameters		
Temperature Range (K)	**Preexponential Factor, A_v (1/s)**	**Activation Energy, E_v (kJ/mol)**
600–830	3282.00	82.42
830–940	32.44	62.48
940–1100	4.48	62.48

Char Oxidation Kinetic Parameters		
Temperature Range (K)	**Preexponential Factor, A_c (kg/ m²s.Pa)**	**Activation Energy, E_c (kJ/mol)**
700–773	8.96E+08	198.16
773–805	1.52E−04	14.09
805–913	4.41E−05	14.09
913–1000	5.33E−03	65.34

the particle remains constant (density of the particle changes) during the experiment. Typical TGA data obtained for char combustion are shown in Figure 3.9. This figure shows three curves: (i) TG curve plotted as the change in the normalized weight of char (%) with respect to temperature, (ii) DTG curve (%/sec), and (iii) simulated TGA data. The char was heated from 700K to 1,173K, and it was

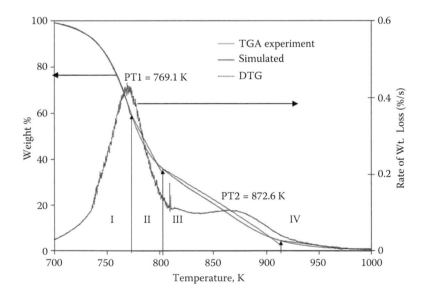

FIGURE 3.9 (See Color Insert.)
Results of char oxidation of a typical high-ash coal (TGA: weight loss; DTG: rate of weight loss).

Moisture : ash : VM : FC :: 12 : 41 : 23 : 24
C : H : N : S : O :: 37 : 2.26 : 0.85 : 0.33 : 6.53

found that above 980K there was negligible change in the weight loss. This temperature of 980K therefore may be considered the burnout temperature (BT) of char. Analysis of the DTG curve for char combustion shows two peak temperatures (PT1 and PT2), as shown in Figure 3.9. The PT1 was observed at 769.1K at a maximum rate weight loss of 0.4394%/s and PT2 at 872.6K at 0.1064%/s.

The TGA model was simulated and fitted to experimental data to obtain kinetic parameters for char oxidation (Figure 3.9). Similar to the devolatilization data, the model could not capture the weight loss profile for char oxidation with a single set of A_c and E_c. Therefore, the whole range was divided into four temperature sub-ranges (I–IV) for fitting the data. The first temperature range was from 700K to 773K, in which almost 40% of the char reacts. The second temperature range was 773K to 805K, where there was around 20% change in weight and this was found to be slower than in the first range. The third sub-range was 800K to 913K, in which more than 30% of the char was oxidized. The remaining 10% char was found to oxidize in the fourth temperature sub-range of 913K to 1,000K. The model parameters in each range are listed in Table 3.4. It can be seen that the kinetic parameters for the first range (showing very fast reactivity of coal) are quite different from the three other temperature sub-ranges. Once the reactive char is oxidized, the rate of reaction slows down and the same can be observed for steps II to IV. Sophisticated optimization tools are available to carry out fitting of the TGA data illustrated here (Deb, 2012; Rao, 2009).

TGA experiments and data processing offer a quick way of characterizing coal samples and estimating the key kinetic parameters of devolatilization and char combustion. TGA experiments and data processing provide insight into the reactivity of fuels and can be useful in quickly comparing them. However, TGA data are limited to rather low values of temperature ramp rate (~1K/s). In real PC fired boilers, coal particles experience much more rapid heating compared to the temperature ramp rates possible in TGA experiments. Very rapid heating of coal particles may alter the various physicochemical changes occurring in the process and thus may influence reactivity and therefore kinetics. The kinetics estimated from TGA experiments therefore may not be directly used while simulating PC fired boilers. It is possible to achieve reasonably high heating rates of particles in a drop-tube furnace (DTF). The DTF experiments and corresponding data processing are therefore expected to provide estimates of kinetics that may be directly used for simulating PC fired boilers. This is discussed in Section 3.3.

3.3 Coal Characterization Using the Drop-Tube Furnace (DTF)

The quality of coal can be estimated based on proximate and ultimate analyses. As discussed earlier, TGA provides a convenient and quick way to assess the devolatilization rates and intrinsic (or chemical) char reactivity at relatively low temperatures (and at low heating rates). TGA avoids the spurious effects due to diffusion limitations and also provides sufficient time for accurate measurements (Russell et al., 1998). The heterogeneity of the maceral content of char, which typically shows two peaks in the DTG plot, can be easily assessed by TGA. However, more often than not, it is essential to characterize coal, especially char combustion, at higher heating rates. DTF experiments are used to obtain such data. Because of the high heating rates offered by DTF, devolatilization is very fast and difficult to characterize via DTF experiments. Devolatilization for fuels that have very high volatile content (e.g., biomass) may however be characterized using the DTF (Jimenez et al., 2008).

The DTF systems are relatively simple and inexpensive (compared with scaled-down combustion systems) and have the ability to mimic key full-scale combustion conditions. In DTF or its variants (e.g., laminar flow reactor or entrained flow reactor), pulverized coal particles and air (with varying concentrations of oxygen) are fed to a tubular reactor (furnace). The particles get heated quickly so that devolatilization is almost instantaneous, and char burning kinetics can be investigated by appropriate data analysis. Feed rates for DTFs range from only 0.1 to 1.0 g coal per minute. Therefore it may not be possible to mimic all aspects of full-scale combustion. However, the DTFs can be configured using key operational parameters to produce results similar to those observed in a full-scale boiler. Key parameters include gas temperature, gas velocity, particle residence time, and excess air. It has short residence times (<5 s), high heating rates (10^4 to 10^5K/s), and varying O_2 concentrations, which represent similar conditions to those in practical systems such as utility boilers.

The char reactivity obtained from TGA experiments is generally the intrinsic (or chemical) reactivity of the char, whereas the kinetic parameters obtained from DTF experiments effectively lump the influence of the internal ash layer diffusion with intrinsic reactivity and hence represent global/apparent kinetic rate parameters. These are particularly useful where coal-related information such as porosity, pore diameter, effectiveness factor, and the surface area of char are not readily available for a particular type of coal.

3.3.1 Typical Example of a DTF Experiment

Typical data obtained from DTF experiments is carbon burnout or conversion recorded at a few specific intervals along the height of the DTF after the coal/char particles are injected into the DTF. These experiments are typically performed at four or five temperatures and also at different outlet oxygen concentrations, and then the carbon conversion data are recorded. After collection, these data are analyzed to obtain kinetic parameters (A_c and E_c) as the oxidation rates are usually represented by Arrhenius-type rate expressions. These estimated kinetic parameters are used as input parameters for combustion submodels in simulating large-scale PC fired boilers that can be based on a lumped model approach (Boyd and Kent, 1986) or more recent computational fluid dynamic (CFD) models (Gupta, 2011; Pallares et al., 2005; Yin et al., 2002; Fan et al., 2001).

Here we illustrate coal characterization with a DTF or entrained flow reactor (EFR) using the experiments performed by Ballester and Jimenez (2005). Ballester and Jimenez (2005) performed experiments using the EFR, consisting of a SiC tube having an internal diameter of 78 mm and total length of 1,600 mm. The SiC tube was heated with the help of electrical heaters placed uniformly around the tube. Pulverized coal along with the primary air was injected from the injection gun at the top of the EFR. The injection gun has the flexibility to vary the height at which the coal particles will be injected into the EFR (up to the upper half part of the EFR). The coal feeder was a rotary valve with a variable frequency drive type of motor to adjust the frequency of rotation of the rotary valve. Uniform and pulsation-free coal feeding was ensured for the experiments. The heated gas was generated using a combustion section with a burner mounted on top of the EFR using natural gas and air. The hot combustion products of the combustor were injected coaxially into the EFR along with the coal and primary air. The set-up also had the flexibility of changing the oxygen concentration at the outlet of the DTF. Solid samples were collected using a sampling probe. The sampling probe and its auxiliary system included a nitrogen-based system for quick quenching of reactions occurring at the surface of the sample and a heated filter to avoid condensation and depositions. This is a typical set-up for DTF and EFR experiments.

Ballester and Jimenez (2005) performed experiments with pulverized anthracite coal samples sourced from the northern Spain region. The proximate and ultimate analyses of coal considered in that work (anthracite) and other operating conditions of the DTF experiments are listed in Table 3.5. The char unburned data at various heights (0.2 m to 1.6 m) were obtained by appropriate combination of injection gun and sampling probe. The unburned fraction was analyzed using TGA.

The processing of DTF data is illustrated in Section 3.3.2 using the experimental data reported by Ballester and Jimenez (2005).

3.3.2 Processing DTF Data

The data obtained from DTF experiments are usually processed using computational models to obtain the parameters of char combustion kinetics. Traditionally, plug flow models are used to represent processes in DTF and for fitting the models to experimental data to obtain the desired kinetic parameters (Tremel and Spliethoff, 2013; Jimenez et al., 2008; Ballester and Jimenez, 2005; Smith, 1973; Field, 1969). The plug flow models are generally simple and can account for the particle size distribution and combustion history of the coal particles along the length of the DTF. Examples of the application of computational models to process DTF data are discussed here.

A plug flow reactor model (no radial variation in gas/solid velocity, temperature, and concentration fields) was used to simulate coal burnout in the

TABLE 3.5

Operating Conditions for DTF Experiments

Burner Inlet	Fuel Air		Heated Gas		
Wall temperature (K)	1,313	1,448	1,573	1,723	1,573
Mass flow rate (kg/s)	5.412×10^{-5} 3.044×10^{-4}	3.044×10^{-4}	3.059×10^{-4}	3.074×10^{-4}	3.141×10^{-4}
Coal flow rate* (kg/s, db)			8.211×10^{-6}		
Inlet stream temperature (K)	298 1,313	1,448	1,573	1,723	1,573
Coal inlet stream temperature (K)	1,313	1,448	1,573	1,723	1,573
Mass fraction of inlet stream					
O_2	0.232 0.0586	0.0586	0.063	0.068	0.112
CO_2	— 0.1149	0.1149	0.114	0.114	0.111
H_2O	0.002 0.094	0.094	0.094	0.093	0.091
O_2 at the outlet of DTF (mole%, db)	4	4	4	4	8

Source: From Ballester, J. and Jimenez, S. (2005). *Combustion and Flame*, 142, 210–222.
Note: *Coal characteristics: Proximate analysis [moisture:ash:volatalies:fixed carbon::1.46:19. 17:10.28:69.09]; Ultimate analysis [C:H:N:S:O::70.3: 3.03:1.63:2.28:2.13]; and HHV = 27.59 MJ/kg.

DTF. The optimized kinetic parameters reported by Ballester and Jimenez (2005) were used in this simulation (Table 3.6). It should be noted that usually it is almost impossible to have a uniform or very narrow particle size distribution of coal for testing. It is therefore essential to account for particle size distribution while processing the DTF data. The particle size distribution (PSD) data were obtained from Ballester and Jimenez (2005) and are

TABLE 3.6

Kinetic Parameters Used with the DTF Model

Devolatilization						Char Oxidation			
Kobayashi et al. (1977)						Ballester and Jimenez (2005)			
BH11 type		VK4 type				Polydispersed		Monodispersed	
A_v (1/s)	E_v (J/kmol)	A_v (1/s)	E_v (J/kmol)	A_v (1/s)	E_v (J/kmol)	A_c (kg/m²-s-Pa)	E_c (J/kmol)	A_c (kg/m²-s-Pa)	E_c (J/kmol)
1.58×10^8	1.29×10^8	1.26×10^7	1.48×10^8	6×10^5	1.44×10^8	0.88×10^{-3}	9.43×10^7	0.4×10^{-3}	$8.3 \times 10^{+7}$

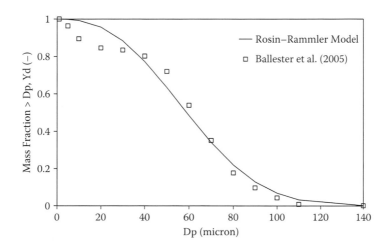

FIGURE 3.10
Rosin-Rammler fit to size distribution of particles used by Ballester and Jimenez (2005).
Mean particle diameter = 68 μm, spread parameter = 2.557

shown in Figure 3.10. The Rosin–Rammler distribution function is based on the assumption that an exponential relationship exists between the particle diameter, d_p, and the mass fraction of particle with diameter greater than d_p as

$$Y_d = \exp(-d_p / \overline{d_p})^n \tag{3.12}$$

where is $\overline{d_p}$ the mean diameter and n is the spread parameter. The Rosin–Rammler equation was fitted to the PSD of the coal particles with a mean particle diameter of 68 μm and a spread parameter of 2.557.

The simulations were carried out using monodispersed as well as polydispersed particles. Simulated results for two operating temperatures (1,313K and 1,723K) are shown in Figure 3.11. The results obtained by assuming monodispersed particles (with a mean diameter D_{43} of 52.2 μm) with corresponding kinetic parameters are also shown in this figure. It can be seen that the model with monodispersed particles over-predicts char burnout. These results confirm the conclusion of Ballester and Jimenez (2005) that char oxidation kinetic parameters should be estimated by considering polydispersed particles rather than monodispersed particles.

However, plug flow models do not account for radial variations in velocity, temperature, O_2 concentration, or the effects of inlet configuration. These may influence the predictions of burnout behavior, as the particles at the same axial distance may experience different oxygen and temperature histories at various radial locations. The particle residence time depends on the particle trajectory it follows after injection. To account for these effects, in recent years, multidimensional computational fluid dynamics (CFD) models have been used to interpret and process data obtained from the DTF.

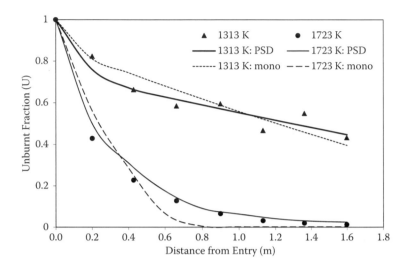

Symbol: Experimental data of Ballester and Jimenez, 2005
Lines: Simulated results

FIGURE 3.11

Plug flow (1D model) simulations for operating temperatures 1,313K and 1,723K.

PSD: Simulations with particle size distribution shown in Figure 3.10
Mono: Simulations with uniform particles of 52.2 μm [D_{43}]

Previous studies are briefly reviewed here before discussing an application of a CFD model to simulate the considered DTF data.

In recent years, substantial progress has been made in the development and application of comprehensive computational combustion models for fossil fuels. These models are now accessible as features in commercially available CFD computer codes. Eaton et al. (1999), in an extensive review paper, discussed the types of data required to validate the predictions of a combustion model and the current status of the combustion models and submodels. Brown et al. (2001) used CFD to model gas and particle flow in a laminar entrained flow reactor (EFR). Their CFD model did not account for reactions in the particles, as they were assumed to be nonreactive. It was also assumed that the impact of solid particles and product gases on the flow and temperature of the bulk flow is negligible. Marklund et al. (2003) investigated an entrained flow gasifier with the fuel injections generating a swirl flow using CFD simulations. The authors highlighted the complexity of the model and the need for validation with the experimental results. Meesri and Moghtaderi (2003) evaluated the capability of a CFD code to predict sawdust combustion conversions. They developed a model that introduced global kinetic parameters obtained through experiments. The agreement between predicted and experimental results was very good. There were also a few attempts to validate the devolatilization or char oxidation characteristics in an EFR using CFD models (Álvarez et al., 2014,

2011; Garba et al., 2013; Ghenai and Janajreh, 2010). Considering these studies and the ability of CFD models to handle complex geometries and simultaneous processes such as fluid flow, heat transfer, particle trajectories, and chemical reactions, CFD models are therefore recommended for analyzing DTF data.

The model equations and boundary conditions were used to simulate the experiments of Ballester and Jimenez (2005); numerical solution of these model equations and simulated results are discussed below.

3.3.2.1 CFD Model Equations

The DTF considered in this work had an internal tube diameter of 78 mm and a total tube length of 1,600 mm. A detailed description of the DTF and experimental procedure can be found in Ballester and Jimenez (2005). There were two inlets: the first one is fuel air (FA) for the injection of transportation/pneumatic air with the coal particles at the center of the drop tube, and second is the coaxial entry for the preheated gas stream (shown schematically in Figure 3.12).

To illustrate the application of the CFD model, a relatively simple two-dimensional (2D) axisymmetric model is considered here. The methodology

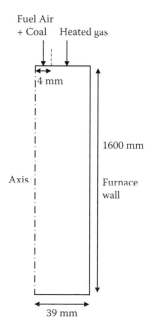

FIGURE 3.12
Two-dimensional axisymmetric domain considered for simulating the drop-tube furnace.

can be extended to a full three-dimensional (3D) model in a straightforward manner. The assumption of a 2D axisymmetric computational domain is adequate for drop-tube furnaces in which flow does not show any significant variation in the azimuthal direction (Sheng et al., 2004). A schematic of the 2D axisymmetric model is shown in Figure 3.12. The model equations were formulated with the following assumptions:

- Drying of coal particles was assumed to be very fast (compared to combustion), and therefore dry coal particles were injected from FA. The moisture present in the coal was added to the fuel air.
- The oxygen first reacts with carbon to form CO, which diffuses into the gas phase and is subsequently oxidized to CO_2.
- The shape of the particle is assumed to be spherical. During the combustion of coal particles, the diameter of the particle remains constant and the density of the particle changes.

Because the discrete phase (coal) was present in low volume fraction (<1%), the Eulerian–Lagrangian approach was used to model two-phase gas-solid flow [see Ranade (2002) for more details on different approaches to model multiphase flows; a more detailed discussion of the Eulerian–Lagrangian approach for modeling gas-solid flows is also included in Chapter 4]. The gas flow in the DTF was laminar (N_{Re} < 500). The particle Reynolds number based on the average particle size of 68 μm is around 0.13 to 0.16.

For a steady-state 2D axisymmetric model, continuity equations for the continuous phase can be written as

$$\frac{1}{r}\frac{\partial}{\partial r}(\rho r U) + \frac{\partial}{\partial z}(\rho W) = S_m \qquad (3.13)$$

where ρ is the density of the fluid, and U and W are the fluid velocity in the radial r and axial direction z, respectively. The S_m is the source term for the total mass added from the discrete phase.

The species conservation equation can be written as

$$\frac{1}{r}\frac{\partial}{\partial r}(\rho r U m_k) + \frac{\partial}{\partial z}(\rho W m_k) = \frac{1}{r}\frac{\partial}{\partial r}\left(\rho D_{km} r \frac{\partial m_k}{\partial r}\right) + \frac{\partial}{\partial z}\left(\rho D_{km}\frac{\partial m_k}{\partial z}\right)$$

$$+ R_k + S_k \qquad (3.14)$$

where m_k is the mass fraction of species k, D_{km} is the diffusion coefficient for species k in the mixture, R_k is the net rate of production of species k by gas-phase chemical reactions, and S_k is the source of species k from the dispersed phase. The net source of chemical species k due to reaction is computed as the sum of the Arrhenius reaction sources over the N_r reactions that the species participate in:

$$R_k = M_{w,k} \sum_{r=1}^{Nr} R_{k,r} \tag{3.15}$$

$$R_{k,r} = \left(v'_{k,r} - v_{k,r}\right) K_r \prod_l \left[C_{lr}\right]^{\eta_{l,r}} \tag{3.16}$$

where $C_{l,r}$ is the molar concentration of each reactant l^{th} species in reaction r, $\eta'_{l,r}$ exponent for each lth reactant in reaction r, $v'_{k,r}$ and $v_{k,r}$ are stoichiometric coefficients for the kth species as product and reactant, respectively, and K_r is the kinetic rate constant.

The source term S_k from the dispersed phase is written as

$$S_k = \frac{\Delta\left(\dot{m}_{pk}\right)}{V} \tag{3.17}$$

$$S_m = \sum_k S_k \tag{3.18}$$

where $\dot{m}_{pk}$ is the particle mass flow rate of component k corresponding to the number of particles that crosses the cell, Δ is the change in the property across that cell, and V is the cell volume. The momentum equation for continuous phase in the radial r and axial z directions can be written as (for steady-state)
Radial momentum, r:

$$\frac{1}{r}\frac{\partial}{\partial r}\left(\rho r UU\right) + \frac{\partial}{\partial z}\left(\rho WU\right) = \frac{1}{r}\frac{\partial}{\partial r}\left(\mu r \frac{\partial U}{\partial r}\right) + \frac{\partial}{\partial z}\left(\mu \frac{\partial U}{\partial z}\right) + F_r + S_U \tag{3.19}$$

Axial momentum, z:

$$\frac{1}{r}\frac{\partial}{\partial r}\left(\rho r UW\right) + \frac{\partial}{\partial z}\left(\rho WW\right) = \frac{1}{r}\frac{\partial}{\partial r}\left(\mu r \frac{\partial W}{\partial r}\right) + \frac{\partial}{\partial z}\left(\mu \frac{\partial W}{\partial z}\right) + F_z + S_W \tag{3.20}$$

Source term for r momentum:

$$S_U = -\frac{\partial p}{\partial r} + \frac{1}{r}\frac{\partial}{\partial r}\left(\mu r \frac{\partial U}{\partial r}\right) + \frac{\partial}{\partial z}\left(\mu \frac{\partial W}{\partial r}\right) - \frac{2\mu U}{r^2} \tag{3.21}$$

Source term for z momentum:

$$S_W = -\frac{\partial p}{\partial z} + \frac{1}{r}\frac{\partial}{\partial r}\left(\mu r \frac{\partial U}{\partial z}\right) + \frac{\partial}{\partial z}\left(\mu \frac{\partial W}{\partial z}\right) \tag{3.22}$$

The momentum source term F for a particular cell is calculated from every particle trajectory j crossing that cell:

$$F_i = \frac{\Delta\left(\dot{m}_{pk}\, u_{p,i}\right)}{V} \tag{3.23}$$

where $u_{p,i}$ are the velocity components of the particle in the ith direction (r or z). The energy balance for gas phase can be written as

$$\frac{1}{r}\frac{\partial}{\partial r}(\rho r U h) + \frac{\partial}{\partial z}(\rho W h) = \frac{1}{r}\frac{\partial}{\partial r}\left(kr\frac{\partial T}{\partial r}\right) + \frac{\partial}{\partial z}\left(k\frac{\partial T}{\partial z}\right) + S_h \tag{3.24}$$

where k is the thermal conductivity of gas and h is an enthalpy. The volumetric source term, S_h, is the sum of the heat of the gas-phase chemical reactions ($S_{h,rxn}$), and the heat from the discrete phase (S_Q) and radiation (S_R).

$$S_h = S_{h,rxn} + S_Q + S_R \tag{3.25}$$

$$h = \sum_k m_k h_k \quad \because h_k = \int_{Tref}^{T} C_{pk}\, dT \tag{3.26}$$

The heat released due to chemical reactions is

$$S_{h,rxn} = -\sum_k \frac{\left(h_k^0 + \int_{Tref}^{T} \Delta C_{pk}\, dT\right) R_k}{M_k} \tag{3.27}$$

where R_k the volumetric rate of creation of species k, and h_k^0 is the formation enthalpy of species k at the reference temperature T_{ref}. M_k is molecular weight of species k.

The heat added from the discrete phase is due to char oxidation:

$$S_Q = \sum_j \frac{\left[(1-f_{heat})\Delta(\dot{m}_c)H_c\right]_j}{V} \tag{3.28}$$

The f_{heat} is the fraction of heat absorbed by the particle, and H_c is the heat released during char oxidation.

The radiative heat transfer in the DTF was modeled using a discrete ordinate (DO) model. The DO model is considered more suitable for systems having an optical thickness (product of characteristic dimension of DTF and absorption coefficient) less than 1 (Sheng et al., 2004). As the optical thickness was found to be less than 0.06 (0.078 [m] * 0.77 [1/m]) for the DTF considered here, the DO model was used to model the radiative heat transfer. The DO model solves the transport equation of radiation intensity I in the direction $\vec{s}$ and can be written as

$$\nabla \cdot (I\vec{s}) + (a + a_p + \sigma_p)I(\vec{r},\vec{s}) = an^2 \frac{\sigma T^4}{\pi} + E_p + \frac{\sigma_p}{4\pi} \int\limits_{\Omega=0}^{4\pi} I(\vec{r},\vec{s}')\phi(\vec{s},\vec{s}')d\Omega' \quad (3.29)$$

where I is the radiant intensity, $\vec{r}$ is a position vector, $\phi(\vec{s},\vec{s}')$ is the scattering phase function, σ is the Boltzmann constant, and a is the absorption coefficient of the gas phase. Here, isotropic scattering (i.e., scattering that is equally likely in all directions) was assumed and for isotropic scattering $\phi(\vec{s},\vec{s}') = 1$. a_p is the equivalent absorption coefficient due to the presence of particulates and is defined as

$$a_p = \lim_{V \to 0} \sum_{n=1}^{N} \varepsilon_{pn} \frac{A_{pn}}{V} \quad (3.30)$$

where $\varepsilon_p n$ and A_{pn} are the emissivity and surface area of particle n, respectively. The equivalent emission E_p is defined as

$$E_p = \lim_{V \to 0} \sum_{n=1}^{N} \varepsilon_{pn} A_{pn} \frac{\sigma T_{pn}^4}{\pi V} \quad (3.31)$$

The equivalent particle scattering factor σ_p is given as

$$\sigma_p = \lim_{V \to 0} \sum_{n=1}^{N} (1 - f_{pn})(1 - \varepsilon_{pn}) \frac{A_{pn}}{V} \quad (3.32)$$

and it is computed during particle tracking. The f_{pn} is the particle scattering factor associated with the nth particle.

The discrete phase was modeled using the Lagrangian approach. The discrete-phase momentum balance on a single particle of size j can be written as

$$\frac{du_{p,i,j}}{dt} = \sum F_{i,j} \quad (3.33)$$

The right-hand side of Equation (3.32) is the sum of the forces acting on the particle of size j in the ith direction. If we consider only gravity and drag force acting on the particle of size j, then we have

$$\sum F_{i,j} = \frac{(\rho_p - \rho_g)}{\rho_p} g + \frac{3}{4} \frac{\rho_g}{\rho_p} \frac{C_D}{d_{p,j}} (u_{p,i,j} - v_i)^2 \qquad (3.34)$$

where ρ_P, d_j, and $u_{p,i,j}$ are the density, diameter, and velocity components, respectively, of the particle of size j in the ith direction (r or z); μ is the viscosity of the gas phase; g is the gravitational constant and C_D is a drag coefficient; and v_i is the velocity component of the gas phase (U or W). The particle Reynolds Number was <0.6. As Morsi and Alexander (1972) correlation cover this range, it was used to calculate C_D.

Different drag laws used in practice are summarized by Ranade (2002) and may be referred to when extending the approach discussed here to different ranges of operation or for nonspherical particles.

The trajectory of particles of size j in the radial and axial directions can be calculated as

$$\frac{dx_{i,j}}{dt} = u_{p,i,j} \qquad (3.35)$$

Species conservation equations for single particles can be written as

$$\frac{d(M_k)}{dt} = S_{pk} \qquad (3.36)$$

The M_k is the mass of species k present in coal.

S_{pk} can be formulated by considering various particle-level phenomena of interest, such as devolatilization and surface reaction–char combustion. Hence, the S_{pk} can be written as

$$S_{pk} = \left[\frac{dM_v}{dt} + \frac{dM_c}{dt} \right] \qquad (3.37)$$

where M_v and M_c are the masses of volatile and char, respectively. As mentioned earlier, drying was assumed to be very fast and therefore does not appear in Equation (3.37). The mass of moisture present in coal is directly added to the source term.

It has been recognized that the single-step models can successfully predict the devolatilization of coal provided that the coal-specific kinetic parameters are known (Jones et al., 2000; Brewster et al., 1995). Hence, the devolatilization was modeled using a simple single-step Arrhenius-type kinetic rate model. The coal devolatilization rate for any particle can be written as (Badzioch and Hawksley, 1970)

$$\frac{dM_v}{dt} = - A_v \, e^{(-Ev/RTp)} M_v \qquad (3.38)$$

where M_v is the mass of volatile present in coal particle at any time, A_v is the preexponential factor, E_v is the activation energy for devolatilization, and T_p is the temperature of the particle.

The char combustion rate was calculated using the kinetic/diffusion model (Baum and Street, 1971; Field, 1969). It was assumed that the char is oxidized to CO by the following reaction:

$$C_{(s)} + 0.5\,O_{2\,(g)} \rightarrow CO_{(g)} \quad \Delta H^\circ = -110\,kJ\,/\,mol$$

This model is simple to implement and needs an apparent kinetic rate constant that accounts for both chemical and internal pore diffusion resistance.

The rate of char oxidation for any particle can be written as (Baum and Street, 1971; Field, 1969):

$$\frac{dM_c}{dt} = -A_p \frac{K_c K_d}{K_c + K_d} P_{O_2} \tag{3.39}$$

Following usual practice, the kinetic rate constant K_c for the char oxidation reaction can be written as

$$K_c = A_C\, e^{(-E_C/RT_P)} \tag{3.40}$$

where A_c is the preexponential factor and E_c is the activation energy for char combustion.

The bulk gas-phase diffusion coefficient K_d for oxidant (Field, 1969) can be given as

$$K_d = \frac{5 \times 10^{-12}}{d_p} \left(\frac{T_g + T_p}{2} \right)^{0.75} \tag{3.41}$$

The unburned fraction of coal U after considering the particle size distribution of coal particles can be written as

$$U = \frac{\sum_j N_{p,j} \left(M_{v,j} + M_{c,j} + M_{w,j} + M_{A,j} \right)}{\sum_j N_{p,j} \left(M_{v0,j} + M_{c0,j} + M_{w0,j} + M_{A0,j} \right)} \tag{3.42}$$

where $M_{v,j}, M_{c,j}, M_{w,j}$ and $M_{A,j}$ are at any time the mass of the volatiles, char, moisture, and ash of particle size j present in the coal; $M_{v0,j}, M_{c0,j}, M_{w0,j}$ and $M_{A0,j}$ are the initial mass of the volatiles, char, moisture, and ash of particle size j present in the coal; and $N_{p,j}$ is the number of particles of size j. The number of particles of size j can be calculated as

$$N_{p,j} = \frac{M^0_{p,j}}{\left(\dfrac{\rho^0_p\, \pi\, \left(d^0_{p,j} \right)^3}{6} \right)} \tag{3.43}$$

where $M_{p,j}^0$ is the initial mass of a single coal particle of size j, ρ_p^0 is the initial density of the coal particle, and $d_{p,j}^0$ is the initial diameter of particle of size j.

The interphase mass source term for any species k present in coal is

$$\Delta\left(\dot{m}_{pk}\right) = \sum_j N_{p,j} \frac{d\left(M_{k,j}\right)}{dt} \tag{3.44}$$

and

$$\Delta\left(\dot{m}_p\right) = \sum_k \Delta\left(\dot{m}_{pk}\right) \tag{3.45}$$

where k represents the species present in coal such as volatiles, char, and moisture.

The volatile material was represented here by single species as $C_{0.1}H_3O_{0.132}$, which is estimated from the proximate and ultimate analyses of coal. The approach presented here, however, can be extended to multiple species, if required, in a straightforward manner.

The following two gas-phase reactions were assumed:

$$CO_{(g)} + 0.5\,O_{2(g)} \rightarrow CO_{2(g)} \quad \Delta H^\circ = -283\ kJ\,/\,mol$$

$$C_{0.1}H_3O_{0.132} + 0.784\,O_{2\,(g)} \rightarrow 0.1\,CO_{2(g)} + 1.5\,H_2O_{(g)} \quad \Delta H^\circ = -262\ kJ\,/\,mol$$

The rate of gas-phase reactions of $C_{0.1}H_3O_{0.132}$ and CO resulting from the char combustion can be represented by an Arrhenius-type rate expression:

$$K_r = A_r\ e^{(-E_r/RT)} \tag{3.46}$$

The energy balance for a single particle can be written as

$$M_p Cp_p \frac{dT_P}{dt} = \left(f_{heat} \frac{dM_c}{dt} H_c \right) + Q_{rad} + Q_{conv} \tag{3.47}$$

Here, M_p is the mass of the particle at any time; Cp_p, $Qrad$, and Q_{conv} are the particle specific heat, and radiative and convective heat transfer, respectively. The particle radiative heat transfer can be written as

$$Q_{rad} = \varepsilon_P \sigma\, A_p \left(T_R^4 - T_P^4\right) \tag{3.48}$$

and the convective heat transfer from the particle can be written as

$$Q_{conv} = hA_P(T_g - T_P) \tag{3.49}$$

TABLE 3.7

Other Parameters Used with the DTF Model

Parameter	Value	Ref.
Particle emissivity (ε_p)	0.9	Backreedy et al. (2006)
Particle scattering factor (f_p)	0.6	Backreedy et al. (2006)
Swelling factor (S_w)	1.0	
Heat fraction (f_{heat})	1.0	Boyd et al. (1986)
Particle density (ρ_p), kg/m³	1700.0	Ballester and Jimenez (2005)
Particle heat capacity (Cp_p), J/kg-K	1700.0	
Emissivity of wall (silicon carbide wall)	0.96	Modest (2003)

where ε_P is the emissivity of the particle, σ is the Stefan–Boltzmann constant ($= 5.67 \times 10^{-8}\ W/m^2K^4$), T_R is the radiation temperature $= (\frac{I}{4\sigma})^{1/4}$, and h is the heat transfer coefficient.

3.3.2.2 Boundary Conditions and Numerical Simulation

The devolatilization kinetic parameters of Kobayashi et al. (1977) of BH11 coal type were used in the model (listed in Table 3.6). Char oxidation kinetic parameters are also listed in Table 3.6.

Two separate inlets were specified for fuel air and heated gas. The pneumatic air and particles were injected from the fuel air inlet, and the coaxial entry of hot gas was introduced from the heated gas inlet (Figure 3.12). The gas flow inlet was defined by the mass flow rate; the outlet was specified as the outlet vent. The gas and particle boundary conditions are specified in Table 3.5. The operating temperature was specified to the wall with emissivity = 0.96 (Modest, 2003). No slip condition was specified to the wall. For a discrete phase, the coal particles were injected from the fuel air inlet by specifying it as group injection. The reflect condition was specified for the particles at the wall, and the escape condition was specified at the outlet. All other model parameters are listed in Table 3.7.

The simulations were performed at various operating temperatures (1,313K; 1,448K; 1,573K; 1,723K) and O_2 concentrations (at the outlet of DTF = 4 mole%, dry basis and = 8 mole%, dry basis). The higher temperature of pneumatic air at the smaller FA injection leads to a high jet velocity (e.g., ~5.3 m/s for 1,723K). Particles were injected at the reactor temperature, and pneumatic air was injected at room temperature. The hot gas temperature was assumed to be the same as the reactor temperature.

Commercial CFD solver FLUENT® (of ANSYS, Inc., Canonsburg, Pennsylvania), version 6.2 was used to solve the mass, energy, and momentum governing equations. The PSD data were obtained from Ballester and Jimenez (2005) and are shown in Figure 3.10. The influence of the number of computational cells was studied by performing simulations on uniform grids of size 5,340 (20 × 267) to 83,148 (78 × 1,066). Based on these results, 20,787 (39 × 533) cells were considered adequate to capture the burnout profile and therefore 20,787 cells were used in all subsequent simulations. Preliminary numerical experiments were carried out to evaluate different discretization schemes and, based on these, a second-order accurate discretization scheme was used for all subsequent simulations. Velocity and pressure coupling was handled by the SIMPLE algorithm. The residuals of velocity components, species, energy, and radiation were monitored. Various criteria, such as an insignificant change (<1%) in velocity, species, temperature, and combustion profiles at various locations of the DTF, were used to decide on the appropriate level of convergence.

3.3.2.3 Application of CFD Models to Simulate DTF Data

The developed CFD model was applied to simulate DTF experiments and to evaluate the sensitivity of predicted results to devolatilization kinetics. Different devolatilization kinetics of similar types of coals (Table 3.5) were used. The simulated results are shown in Figure 3.13. It can be seen that although the amount of volatiles present in coal was small (10.3 wt%), the model predictions were sensitive to devolatilization kinetic parameters. The kinetic parameters reported by Ballester and Jimenez (2005) showed a convex nature up to a 0.6-m distance and did not show the trends observed in experiments. Simulated results with the kinetic parameters reported by Kobayashi et al. (1977) captured the trends observed in the experimental results. Therefore, for further simulations, the devolatilization kinetic parameters reported for BH-11 coal by Kobayashi et al. (1977) were used.

Initially, the kinetic parameters obtained from the plug flow model were used in the 2D axisymmetric CFD model. The results for two operating temperatures (1,313K and 1,573K) are shown in Figure 3.14. It can be seen that these parameters significantly under-predict the combustion profile. The sensitivity of simulated results to values of the preexponential factor (A_c) is shown in Figure 3.15.

The simulated results indicate that it may be possible to obtain multiple pairs of preexponential factor (A_c) and activation energy (E_c) to represent the DTF experimental data. Simulated results with three different pairs of these parameters are shown in Figure 3.16 for 1,573K. It can be seen that there is no significant difference in simulated burnout profiles with these three pairs of parameters.

Any of these pairs can therefore be used for subsequent evaluation. The simulated burnout profiles for four operating temperatures (1,313K; 1,448K; 1,573K; 1,723K) are shown in Figure 3.17. The following kinetic parameters were used for carrying out these simulations:

Char oxidation kinetic parameters: $A_c = 2.7 \times 10^{-3}$ kg/m²-s-Pa and $E_c = 9.43$ kJ/mol

Devolatilization kinetic parameters: $A_v = 1.58 \times 10^8$ 1/s and $E_v = 1.29$ kJ/mol

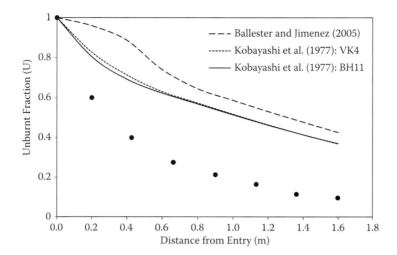

Symbol: Experimental data of Ballester and Jimenez, (2005)
Lines: Simulated results based on kinetic parameters listed in Table 3.6

FIGURE 3.13
Sensitivity of simulated coal burnout to devolatilization kinetics.

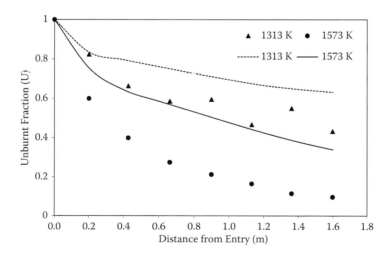

Symbol: Experimental data of Ballester and Jimenez, (2005)
Lines: Simulated results

FIGURE 3.14
Simulated profiles of coal burnout for $A_c' = 0.88 \times 10^{-3}$ kg/m²-s-Pa and $\bar{E}_c = 94$ kJ/mol.

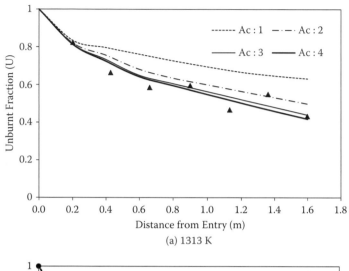

(a) 1313 K

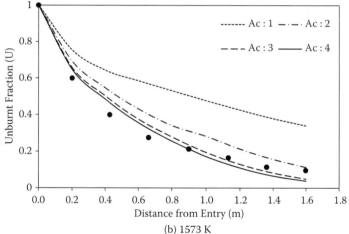

(b) 1573 K

$$Ac:1 \equiv 8.8 \times 10^{-4} \frac{kg}{m^2 sPa} \quad 2 \equiv 2.0 \times 10^{-3} \frac{kg}{m^2 sPa} \quad 3 \equiv 2.5 \times 10^{-3} \frac{kg}{m^2 sPa} \quad 4 \equiv 2.7 \times 10^{-3} \frac{kg}{m^2 sPa}$$

Symbols : Experimental data of Ballester and Jimenez (2005)

Lines : Simulated results (E_c = 9.43 kJmol)

FIGURE 3.15
Sensitivity of simulated coal burnout profile to A_c.

Simulations were also performed at different oxygen concentrations to gain an understanding of the applicability of these evaluated parameters at various operating conditions. The simulated results with two different operating conditions (adjusted such that the oxygen concentration at the outlet was 4% in one case and 8% in the other case) are shown in Figure 3.18. It can be seen from Figures 3.17 and 3.18 that the model was reasonably able to capture the observed combustion profiles at these operating conditions.

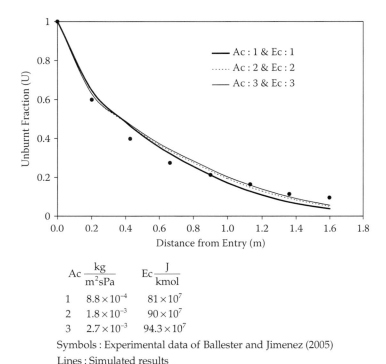

FIGURE 3.16
Simulated coal burnout profiles for three sets of char oxidation kinetics (at 1,573K).

Sophisticated optimization algorithms can be used (unlike the ad hoc trial-and-error method presented here for the sake of illustration) for improving the fit and identifying optimal kinetic parameters from the DTF data.

To highlight the differences in between the plug flow and 2D axisymmetric CFD models, the contour plots of simulated coal burnout profiles with plug flow and 2D axisymmetric models are shown in Figure 3.19. It can be seen from this figure that a significant fraction of the injected particles travel along the axis after injection with minimal lateral dispersion of the particles. Consequently, particles travel with a much higher velocity (see contours shown in Figure 3.19(a) indicating that the gas velocity along the axis is around 0.8 to 0.9 m/s and around 0.1 m/s near the wall region) compared to that with the plug flow model (with an average velocity of ~0.38 m/s; see Figure 3.19(b)).

The residence time distributions (RTDs) of particles obtained from 2D axisymmetric and plug flow simulations (1D model) are shown in Figure 3.20. The calculated values of mean residence time for the plug flow model and 2D CFD model were 3.87 s and 1.82 s, respectively. The mean residence time of the particles for the 2D CFD model was smaller than the 1D model; the 2D

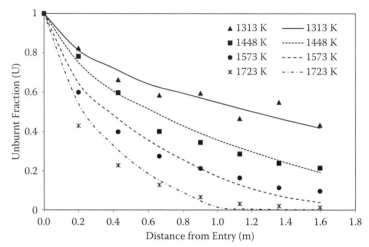

Symbols : Experimental data of Ballester and Jimenez (2005)
Lines : Simulated results

Devolatilization kinetic parameters : $A_v = 1.58 \times 10^8 \dfrac{1}{s}$, $E_v = 1.29 \times 10^8 \dfrac{J}{kmol}$

Char oxidation kinetic parameters : $A_c = 2.7 \times 10^{-3} \dfrac{kg}{m^2 s Pa}$, $E_c = 9.43 \times 10^7 \dfrac{J}{kmol}$

FIGURE 3.17
Simulated coal burnout profiles at different operating temperatures.

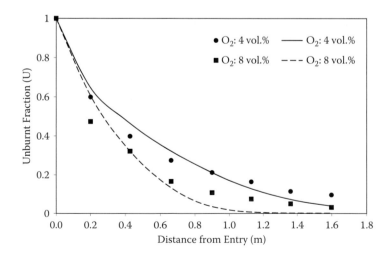

Symbol: Experimental data of Ballester and Jimenez, (2005)
Lines: Simulated results
Oxygen concentration mentioned in legend indicates volume percent of oxygen at the outlet
of DTF.

FIGURE 3.18
Simulated coal burnout profiles at different oxygen concentrations (at 1,573K).

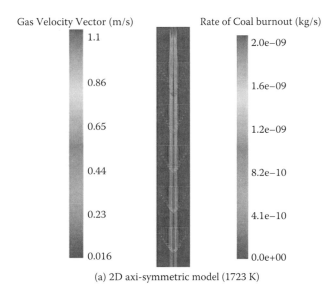

(a) 2D axi-symmetric model (1723 K)

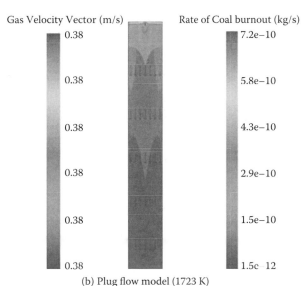

(b) Plug flow model (1723 K)

FIGURE 3.19 (See Color Insert.)
Distribution of char burnout rates (superimposed with velocity magnitude vectors): (a) 2D axisymmetric model (1,723K) and (b) plug flow model (1,723K).

CFD model under-predicted the combustion profile when the kinetic parameters based on the plug flow model were implemented. Therefore, it is recommended to use at least 2D axisymmetric CFD models to estimate kinetic parameters from the DTF experiments. The influence of inlet configuration on the flow profile is significant in the DTF, and adequate care must be taken

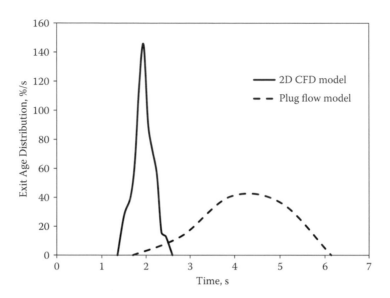

FIGURE 3.20
Residence time distribution (RTD) of coal particles (T = 1,723K) as estimated by CFD models.

to include those influences in the model used for estimating kinetic parameters. Sometimes, swirling inlets are used with the DTF and, in such cases,
it may be necessary to use full 3D models to process DTF experimental data
for estimating appropriate kinetic parameters.

3.4 Summary and Conclusions

Coal is a heterogeneous mixture of organic material, moisture, and mineral
matter. The proximate, ultimate, and ash analyses provide the primary characterization of coal. This is complemented with the estimation of gross calorific value (GCV). The proximate analysis (quantification of volatiles, fixed
char, moisture, and ash) and the ultimate analysis (carbon, hydrogen, oxygen,
nitrogen, and sulfur (CHONS) contents) do not provide information about the
reactivity of the coal. Coal reactivity depends on devolatilization, char struc-
tures, and the different types of microscopic components. Thermogravimetric
analysis (TGA) and the drop-tube furnace (DTF) offer convenient and effective
platforms for characterizing coal. TGA is mainly used to characterize devolatilization, and the DTF is mainly used for char combustion kinetics. Some of
the key points discussed in this chapter are summarized here:

- Single reaction models (SRMs) are often used to fit the TGA data in
  order to estimate the devolatilization kinetic parameters. It may be

necessary to divide the overall temperature range into multiple sub-ranges to improve the quality of fit with SRMs.

- Multiple reaction models (MRMs) such as distributed activation energy models (DAEMs) provide a better framework than SRMs to represent TGA data. However, implementation of these DAEMs into a commercial CFD framework is more complicated than that for SRMs.

- TGA offers modest heating rates (up to 1K/s). TGA is suitable for quickly characterizing and comparing different fuels (coals). It is also suitable for estimating devolatalization kinetics. However, the heating rates in real PC fired boilers are orders of magnitude higher than those that can be realized in TGA.

- The DTF offers much higher heating rates compared to TGA and therefore offers conditions more relevant to PC fired boilers. The DTF is therefore suitable for characterizing char combustion kinetics.

- It is recommended to use multidimensional CFD models to interpret DTF data for estimating char combustion kinetics.

Modeling of TGA and DTF experiments and data not only provides useful information and understanding about fuels under consideration, but also provides an effective basis for undertaking computational modeling of full-scale PC fired boilers. The CFD models and reactor network models for PC boilers are discussed in Chapter 4 and 5, respectively.

References

Aboyade, A.O., Carrier, M., Meyer, E.L., Knoetze, J.H., and Gorgens, J.F. (2012). Model fitting kinetic analysis and characterisation of the devolatilization of coal blends with corn and sugarcane residues, *Thermochim. Acta*, 530, 95–106.

Alonso, M.J.G., Borrego, A.G., Alvarez, D., Kalkreuth, W., and Menéndez, R. (2001). Physicochemical transformations of coal particles during pyrolysis and combustion, *Fuel*, 80, 1857–1870.

Álvarez, L., Gharebaghi, M., Pourkashanian, M., Williams, A., Riaza, J., Pevida, C., and Rubiera, F. (2011). CFD modelling of oxy-coal combustion in an entrained flow reactor, *Fuel Processing Technol.*, 92(8), 1489–1497.

Álvarez, L., Yin, C., Riaza, J., Pevida, C., Pis, J.J., and Rubiera, F. (2014). Biomass co-firing under oxy-fuel conditions: A computational fluid dynamics modelling study and experimental validation, *Fuel Processing Technol.*, 120, 22–33.

Backreedy, R.I., Fletcher, L.M., Ma, L., Pourkashanian, M., and Williams, A. (2006). Modelling pulverized coal combustion using a detailed coal combustion model, *Combust. Sci. and Technol.*, 178, 763–787.

Badzioch, S. and Hawksley, P.G.W. (1970). Kinetics of thermal decomposition of pulverized coal particles, *Ind. Eng. Chem. Process Design and Development*, 9, 521–530.

Ballester, J. and Jimenez, S. (2005). Kinetic parameters for the oxidation of pulverized coal as measured from drop tube tests, *Combustion and Flame*, 142, 210–222.

Baum, M.M. and Street P.J. (1971). Predicting the combustion behavior of coal particles, *Combustion Sci. Technol.*, 3, 231–243.

Boyd, R.K. and Kent, J.H. (1986). Three-dimensional furnace computer modeling, In *21st Symp. (Int.) on Combustion*, The Combustion Institute, pp. 265–274

Brewster, B.S., Smoot, L.D., Barthelson, S.H., and Thornock, D.E. (1995). Model comparisons with drop tube combustion data for various devolatilization sub models, *Energy & Fuels*, 9, 870–879.

Brown, A.L., Dayton, D.C., Nimlos, M.R., and Daily, J.W. (2001). Design and characterization of an entrained flow reactor for the study of biomass pyrolysis chemistry at high heating rates, *Energy & Fuels*, 15(5), 1276–1285.

Caballero, J.A. and Conesa, J.A. (2005). Mathematical considerations for nonisothermal kinetics in thermal decomposition, *J. Anal. Appl. Pyrolysis*, 73, 85–100.

Cai, J. and Ji, L. (2007). Pattern search method for determination of DAEM kinetic parameters from nonisothermal TGA data of biomass, *J. Math. Chem.*, 42(3), 547–553.

Cai, J., Li, T., and Liu, R. (2011). A critical study of the Miura–Maki integral method for the estimation of the kinetic parameters of the distributed activation energy model, *Bioresource Technol.*, 102, 3894–3899.

Caprariis, B.D., Fillippis, P.D., Herce C., and Verdone, N. (2012). Double-Gaussian distributed activation energy model for coal devolatilization, *Energy & Fuel,.* 26(10), 6153–6159.

Card, J.B.A. and Jones, A.R. (1995). A drop tube furnace study of coal combustion and unburned carbon content using optical techniques, *Combustion and Flame*, 101, 539–547.

Chandra, A. (2009). Enhancement of collection efficiencies of electrostatic precipitators: Indian experiments, *11th Int. Conf. Electrostatic Precipitation, Hangzhou,* 2008, pp. 27–34. (Editor: Keping Yan, Co-published by Zhejiang University Press, Hangzhou and Springer-Verlag GmbH, Berlin Heidelberg).

Chen, W.-H. and Wu, J.-S. (2009). An evaluation on rice husks and pulverized coal blends using a drop tube furnace and a thermogravimetric analyzer for application to a blast furnace, *Energy*, 34(10), 1458–1466.

Chi, T., Zhang, H., Yan, Y., Zhou, H., and Zheng, H. (2010). Investigations into the ignition behaviors of pulverized coals and coal blends in a drop tube furnace using flame monitoring techniques, *Fuel*, 89(3), 743–751.

Cloke, M., Lester, E., and Thompson, A.W. (2002). Combustion characteristics of coals using a drop-tube furnace, *Fuel*, 81, 727–735.

Cloke, M., Wu, T., Barranco, R., and Lester, E. (2003). Char characterisation and its application in a coal burnout model, *Fuel*, 82, 1989–2000.

Cumming, J.W. (1984). Reactivity assessment of coals via a weighted mean activation energy, *Fuel*, 63, 1436–1440.

Damartzis, T., Vamvuka, D., Sfakiotakis, S., and Zabaniotou, A. (2011). Thermal degradation studies and kinetic modeling of cardoon (*Cynara cardunculus*) pyrolysis using thermogravimetric analysis (TGA), *Bioresource Technol.*, 102(10), 6230–6238.

Deb, K. (2012). *Optimization for Engineering Design: Algorithms and Examples, 2nd edition*, Prentice-Hall of India PVT. Ltd.

Diessel, C.F.K. (1992). *Coal-Bearing Depositional Systems*, Springer-Verlag, Berlin.

Eaton, A.M., Smoot, L.D., Hill, S.C., and Eatough, C.N.(1999). Components, formulations, solutions, evaluation, and application of comprehensive combustion models, *Progr. Energy and Combustion Sci.*, 25, 387–436.

Essenhigh, R.H. and Mescher, A.M. (1996). Influence of pressure on the combustion rate of carbon, *Symp. (Int.) on Combustion*, 26, 3085.

Essenhigh, R.H., Klimesh, H.E., and Fortsch, D. (1999). Combustion characteristics of carbon: Dependence of the zone i-zone ii transition temperature (tc) on particle radius, *Energy & Fuels*, 13, 826–831.

Fan, J., Qian, L., Ma, Y., Sun, P., and Cen, K. (2001). Computational modeling of pulverized coal combustion processes in tangentially fired furnaces, *Chem. Eng. J.*, 81(1), 261–269.

Farrow, T.S., Sun, C., and Snape, C.E. (2013). Impact of biomass char on coal char burn-out under air and oxy-fuel conditions, *Fuel*, 114, 128–134.

Field, M.A. (1969). Rate of combustion of size graded fractions of char from a low-rank coal between 1200K and 2000K, *Combustion and Flame*, 13, 237–252

Fiori, L., Valbusa, M., Lorenzi, D., and Fambri, L.(2012). Modeling of the devolatilization kinetics during pyrolysis of grape residues, *Bioresource Technol.*, 103, 389–397.

Fletcher, T.H. and Kerstein, A.R. (1992). Chemical percolation model for devolatilization: 3. Direct use of 13C NMR data to predict effects of coal types, *Energy & Fuels*, 6, 414.

Garba, M.U., Ingham, D.B., Ma, L., Degereji, M.U., Pourkashanian, M., and Williams, A. (2013). Modelling of deposit formation and sintering for the co-combustion of coal with biomass, *Fuel*, 113, 863–872.

Gašparoviè, L., Labovský, J., Markoš, J., and Jelemenský, L., (2012). Calculation of kinetic parameters of the thermal decomposition of wood by distributed activation energy model (DAEM), *Chem. Biochem. Eng.*, 26(1), 45–53.

Gavalas, G.R., (1982). *Coal Pyrolysis*, Elsevier, Amsterdam

Ghenai, C. and Janajreh, I. (2010). CFD analysis of the effects of co-firing biomass with coal, *Energy Conversion and Management*, 51(8), 1694–1701

Gupta, R.P., Yan, L., Kennedy, E.M., Wall, T.F., Masson, M., and Kerrison, K. (1999). System accuracy for CCSEM analysis of minerals in coal. In Gupta, R.P., Wall, T.F., Baxter, L. (Eds.), *Impact of Mineral Impurities in Solid Fuel Combustion*. Kluwer Academic/Plenum Publishers, New York, pp. 225–235.

Gupta, D.F. (2011). Modeling of Coal Fired Boiler, PhD thesis, University of Pune, India.

Hampartsoumian, E., Nimmo, W., Rosenberg, P., Thomsen, E., and Williams, A. (1998). Evaluation of the chemical properties of coals and their maceral group constituents in relation to combustion reactivity using multi-variate analyses, *Fuel*, 77(7), 735–748.

Haykiri-Acma, H., Baykan, A., Yaman, S., and Kucukbayrak, S. (2013). Are medium range temperatures in drop tube furnace really ineffective? *Fuel*, 105, 338–344.

Hindmarsh, C.J., Thomas, K.M., Wang, W.X., Cai, H.Y., Güell, A.J., Dugwell, D.R., and Kandiyoti, R. (1995). A comparison of the pyrolysis of coal in wire-mesh and entrained-flow reactors, *Fuel*, 74, 1185–1190.

Howard, J.B., Fong, W.S., and Peters, W.A. (1987). In *Fundamentals of the Physical Chemistry of Pulverized Coal Combustion, NATO ASI Series, Series E: Applied Sciences*, No. 137 (J. Layaye and G. Prado, Eds.), Martinus Nijhoff, Netherlands, p. 77.

Hurt, R., Sun, J., and Lunden M. (1998). A kinetic model of carbon burnout in pulverized coal combustion, *Combustion & Flame*, 113, 181–197.

Hurt, R.H. and Gibbins, J.R. (1995). Residual carbon from pulverized coal fired boilers. 1. Size distribution and combustion reactivity, *Fuel*, 74, 471–480.

Jain, A.A., Mehra, A., and Ranade, V.V. (2014). Processing of TGA data for Indian coal, unpublished work.

Jimenez, S., Remacha, P., Ballesteros, J.C., Gimenez, A., and Ballester, J. (2008). Kinetics of devolatilization and oxidation of a pulverized biomass in an entrained flow reactor under realistic combustion conditions, *Combustion & Flame*, 152(4), 588–603.

Jones, J.M., Kubacki, M., Kubica, K., Ross, A.B., and Williams, A. (2005). Devolatilisation characteristics of coal and biomass blends. *J. Anal. Appl. Pyrol.*, 74(1-2), 502–511.

Jones, J.M., Patterson P.M., Pourkashanian, M., Williams, A., Arenillas, A., Rubiera, F., and Pis, J.J. (1999). Modeling NOx formation in coal particle combustion at high temperature: An investigation of the devolatilization kinetic factors, *Fuel*, 78, 1171–1179.

Jones, J.M., Pourkashanian, M., Williams, A., and Hainsworth, D. (2000). A comprehensive biomass combustion model, *Renewable Energy*, 19(1), 229–234.

Jüntgen, H. and Van Heek, K.H. (1979). An update of German non-isothermal coal pyrolysis work, *Fuel Processing Technol.*, 2, 261–293.

Kobayashi, H., Howard, J.B., and Sarofim, A.F. (1977), Coal devolatilization at high temperatures, *Proc. Combust. Inst.*, 16, 411–425.

Kumar, M. and Patel, S.K. (2008). Characteristics of Indian non-coking coals and iron ore reduction by their chars for directly reduced iron production, *Mineral Processing and Extractive Metallurgy Review: An International Journal*, 29 (3), 258–273.

Li, S., Chen, X., Wang, L., Liu, A., and Yu, G. (2013). Co-pyrolysis behaviors of saw dust and Shenfu coal in drop tube furnace and fixed bed reactor, *Bioresource Technol.*, 148, 24–29.

Marklund, M., Gebart, R., and Fletchet, D.F. (2003). In *Proc. Colloquium on Black Liquor Combustion and Gasification*. Park City, Utah, USA, May 2003.

Meesri, C. and Moghtaderi, B. (2003). Experimental and numerical analysis of sawdust-char combustion reactivity in a drop tube reactor, *Combust. Sci. and Technol.*, 175(4), 793–823.

Mishra A. (2009). Assessment of Coal Quality of Some Indian Coals, B. Tech. Thesis, Department of Mining Engineering, National Institute of Technology, Rourkela.

Miura, K. (1995). A new and simple method to estimate $f_{(E)}$ and $k_0^{(E)}$ in the distributed activation energy model from three sets of experimental data, *Energy & Fuels*, 9(2), 302–307.

Miura, K. and Maki, T. (1998). A simple method for estimating $f(E)$ and ko(E) in the distributed activation energy model, *Energy & Fuels*, 12(5), 864–869.

Modest, M.F. (2003). *Radiative Heat Transfer, 2nd edition*, Academic Press, Burlington, MA, USA.

Morgan, P.A., Robertson, S.D., and Unsworth, J.F. (1986). Combustion studies by thermogravimetric analysis. 1. Coal oxidation, *Fuel*, 65, 1546–1551.

Morsi, S.A. and Alexander, A.J. (1972). An investigation of particle trajectories in two-phase flow systems, *J. Fluid Mech.*, 55(2), 193–208.

Niksa, S. and Kerstein, A.R. (1991). FLASHCHAIN theory for rapid coal devolatilization kinetics. 1. Formulation, *Energy& Fuels*, 5(5), 647–665

Pallares, J., Arauzo, I., and Diez, L.I. (2005). Numerical prediction of unburned carbon levels in large pulverized coal utility boilers, *Fuel*, 84, 2364–2371.

Parikh, J., Channiwala, S.A., and Ghosal, G.K. (2005). A correlation for calculating HHV from proximate analysis of solid fuels, *Fuel*, 84, 487–494.

Ranade, V.V. (2002). *Computational Flow Modeling for Chemical Reactor Engineering*, Academic Press, London.

Rao, S.S. (2009). *Engineering Optimization: Theory and Practice, fourth edition*, John Wiley & Sons, Inc., New York.

Russell, N.V., Beeley, T.J., Man, C.K., Gibbins, J.R., and Williamson, J. (1998). Development of TG measurements of intrinsic char combustion reactivity for industrial and research purpose, *Fuel Processing Technol.*, 57(2), 113–130.

Sahu, S.G., Sarkar, P., Chakraborty, N., and Adak, A.K. (2010). Thermogravimetric assessment of combustion characteristics of blends of a coal with different biomass chars, *Fuel Processing Technol.*, 91(3), 369–378.

Sarkar, P., Sahu, S.G., Mukherjee, A., Kumar, M., Adak, A.K., Chakraborty, N., and Biswas, S. (2014). Co-combustion studies for potential application of sawdust or its low temperature char as co-fuel with coal, *Appl. Thermal Eng.*, 63(2), 616–623.

Sarkar, S. (2009). *Fuels and Combustion, 3rd edition*, Universities Press Pvt. Ltd., India.

Seo, D.K., Park, S., Kim, Y.T., Hwang, J., and Yu, T.-U. (2011). Study of coal pyrolysis by thermo-gravimetric analysis (TGA) and concentration measurements of the evolved species, *J. Anal. Appl. Pyrolysis*, 92(1), 209–216.

Serio, M.A., Hamblen, D.G., Markham, J.R., and Solomon, P.R. (1987). Kinetics of volatile product evolution in coal pyrolysis: Experiment and theory, *Energy & Fuels*, 1(2), 138–152.

Sheng, C., Moghtaderi, B., Gupta, R., and Wall, T.F. (2004). A computational fluid dynamics based study of the combustion characteristics of coal blends in pulverized coal-fired furnace, *Fuel*, 83, 1543–1552.

Smith, I.W. (1982). The combustion rates of coal chars: A review, *Symp. (Int.) on Combustion*, 19, 1045–1065.

Solomon, P.R., Hamblen, D.G., Carangelo, R.M., Serio, M.A., and Deshpande, G.V. (1988). General model of coal devolatilization, *Energy & Fuels*, 2(4), 405–422.

Solomon, P.R. and Fletcher, T.H. (1994). Impact of coal pyrolysis on combustion, *Symp. (Int.) on Combustion*, 25, 463–474.

Solomon, P.R., Serio, M.A., and Suuberg, E.M. (1992). Coal pyrolysis: Experiments, kinetic rates and mechanisms, *Progr. Energy and Combustion Sci.*, 18(2), 133–220.

Speight, J.G. (2013). *The Chemistry and Technology of Coal*, Third Edition, CRC Press, Taylor & Francis Group, Boca Raton, FL, USA.

Sun, J.K. and Hurt, R.H. (2000). Mechanisms of extinction and near-extinction in pulverized solid fuel combustion, *Proc. Combustion Institute*, 28(2), 2205–2213.

Sun, J., Hurt, R.H., Niksa, S., Muzio, L., Mehta, A., and Stallings, J. (2003). A simple numerical model to estimate the effect of coal selection on pulverized fuel burnout, *Combust. Sci. and Technol.*, 175, 1085–1108.

Tremel, A. and Spliethoff, H. (2013). Gasification kinetics during entrained flow gasification. Part III. Modelling and optimisation of entrained flow gasifiers, *Fuel*, 107, 170–182.

Vleeskens, J.M. and Nandi, B.N. (1986). Burnout of coals: Comparative bench-scale experiments on pulverized fuel and fluidized bed combustion, *Fuel*, 65, 797–802.

Wang, G., Silva, R.B., Azevedo, J.L.T., Martins-Dias, S., and Costa, M. (2014b). Evaluation of the combustion behaviour and ash characteristics of biomass waste derived fuels, pine and coal in a drop tube furnace, *Fuel*, 117, 809–824.

Wang, G., Zander, R., and Costa, M. (2014a). Oxy-fuel combustion characteristics of pulverized-coal in a drop tube furnace, *Fuel*, 115, 452–460.

Williams, A., Pourkashanian, M., and Jones, J.M. (2000). The combustion of coal and some other solid fuels, *Proc. Combustion Institute*, 28(2), 2141–2162.

Yin, C., Caillat, S., Harion, J., Baudoin, B., and Perez, E. (2002). Investigation of the flow, combustion, heat-transfer and emissions from a 609 MW utility tangentially fired pulverized-coal, *Fuel*, 81(8), 997–1006.

Yoshizawa, N., Maruyama, K., Yamashita, T., and Akimoto, A. (2006). Dependence of microscopic structure and swelling property of DTF chars upon heat-treatment temperature, *Fuel*, 85, 2064–2070.

Yuzbasi, N.S. and Selçuk, N. (2011). Air and oxy-fuel combustion characteristics of biomass/lignite blends in TGA-FTIR, *Fuel Processing Technol.*, 92(5), 1101–1108.

Zhang, Y., Wei, X.-L., Zhou, L.-X., and Sheng, H.-Z. (2005). Simulation of coal combustion by AUSM turbulence-chemistry char combustion model and a full two-fluid model, *Fuel*, 84, 1798–1804.

Zolin, A., Jensen, A., and Dam-Johansen, K. (2001). Coupling thermal deactivation with oxidation for predicting the combustion of a solid fuel, *Combustion and Flame*, 125, 1341–1360.

4

CFD Model of a Pulverized
Coal Fired Boiler

The pulverized coal (PC) fired boiler is a key piece of equipment governing the overall energy efficiency of coal fired power stations. The performance of a PC fired boiler depends on several design and operating parameters, such as configuration of the boiler; the number, location, and design of burners (tilt and swirl at burners); the locations and configurations of internal heat exchangers; coal characteristics (quality (coal rank), sulfur and nitrogen contents, particle size distribution, composition, reactivity and its feed rate); extent of excess air; possibility of air ingress; imbalance in temperatures of steam drums; heat transfer effectiveness, etc. It is essential to develop a comprehensive understanding of the influence of furnace configuration, burner design, and different operating parameters on the overall performance of a coal fired boiler. Information on the temperature field within the boiler and local heat transfer coefficients at the boiler tubes is of interest. Knowledge of particle trajectories (bottom ash as well as fly ash) is also one of the key interests in understanding the long-term performance of coal fired boilers. The particles may interact with preheater and superheater tubes. Understanding such interactions is important in estimating erosion rates. Because many of these parameters are strongly coupled and may influence performance in a complex way, it is always desirable to develop a computational model that establishes a relationship between hardware design/operating parameters and PC boiler performance.

To establish quantitative relationships between hardware/operating parameters and performance, it is necessary to adequately model these various processes that are occurring in the PC fired boiler. Models need to adequately account for various key issues, namely turbulent flow and transport processes, the motion of coal particles, devolatilization, burning of char, combustion of volatile components, radiative heat transfer, etc. A computational fluid dynamics (CFD) framework offers such a possibility. CFD has evolved as a powerful design and predictive tool in recent years to simulate complex industrial process equipment (Ranade, 2002). It has been successfully used to simulate large utility boilers (Gupta, 2011; Belosevic et al., 2006; Pallares et al., 2005; Yin et al., 2002; Fan et al., 2001; Eaton et al., 1999; Boyd and Kent, 1986). The CFD models represent processes occurring inside PC fired boilers with the help of mass, momentum, and energy conservation equations. A systematic methodology must be developed to obtain useful results from CFD models and also for using these results to realize performance enhancement in practice. The development and solution of CFD models and submodels (such as combustion and heat transfer) for simulating PC boilers are discussed in this chapter.

4.1 Formulation of CFD Model of a PC Fired Boiler

CFD is a body of knowledge and techniques to solve mathematical models of flow processes on digital computers. Any rigorous analysis of flow processes starts with the application of the universal laws of conservation of mass, momentum, and energy. It is very important to clearly identify and understand the implications of the underlying assumptions (both explicit and implicit) while describing physical processes in mathematical equations. Section 4.1.1 discusses the basic governing equations describing the flow, heat transfer, and reactions occurring in PC fired boilers. Subsequent sections discuss various strategies for applying these equations in simulating PC fired boilers and illustrating the application of the CFD model to a typical 210-MWe PC fired boiler.

The basic governing equations based on the three conservation (mass, momentum, and energy) laws should be complemented by relevant constitutive equations and equations of state for the fluids under consideration to close the system of equations. There are two approaches for deriving basic governing equations: In the Eulerian approach, an arbitrary control volume in a stationary reference frame is used for deriving the basic governing equations. In an alternative Lagrangian approach, equations are derived by considering a control volume (material volume) such that the velocity of the control volume surface always equals the local fluid velocity. For single-phase flows, both approaches give the same final form of conservation equations (see Ranade (2002) for more details). These two approaches, however, offer different routes to simulate multiphase flow processes.

The Eulerian–Lagrangian approach is usually used and recommended for modeling dilute gas-solid flows in PC fired boilers. In the Eulerian–Lagrangian approach, the motion of the continuous phase is modeled using the Eulerian framework, and the motion of dispersed-phase particles is modeled using the Lagrangian framework. Averaging over a large number of particle trajectories is then carried out to derive the required information for modeling the continuous phase. In this approach, particle-level processes such as reactions, heat and mass transfer, etc. can be simulated in adequate detail. In the case of turbulent flows, it is necessary to simulate very large numbers of particle trajectories to obtain meaningful averages. There have been some attempts to use the Eulerian–Eulerian approach for modeling a PC fired boiler (Li et al., 2003; Zhou et al., 2002; Guo and Chan, 2000). However, considering the low volume fraction of dispersed phase, we recommend the Eulerian–Lagrangian approach for simulating PC fired boilers and therefore the Eulerian–Eulerian approach is not discussed here. Details of individual modeling aspects of the Eulerian–Lagrangian approach are discussed in the following, along with the basic governing equations.

The flow in PC fired boilers is inherently turbulent. Turbulent flows span a large range of spatiotemporal scales. The demands on computational resources to resolve all the relevant time and space scales of turbulent motion may push the computations of turbulent flows in large industrial equipment such as PC fired boilers beyond the realm of present computing capabilities. Estimates from various sources differ as to when computer technology will advance to the point where turbulent flow calculations can be performed from first principles. It appears that most engineering computations of turbulent flow processes will have to rely on models for turbulent flows, at least for the foreseeable future. This is especially true for PC fired boiler-type applications, where, in addition to turbulence, there are many other complexities (e.g., chemical reactions, multiple phases, radiation, complex geometry, etc.).

Major efforts in the area of turbulence modeling were and are still directed toward developing tractable computational models for turbulent flows with reasonable demands on computational resources. A large number of models have been developed in the past few decades. These modeling approaches can be classified into three categories: (1) direct numerical simulations (DNS), (2) large eddy simulations (LES), and (3) Reynolds averaged Navier–Stokes (RANS) equations. As one progresses from DNS to RANS, more and more of the turbulent motion is approximated and, therefore, requires fewer computational resources. Flow and other processes occurring in a PC fired boiler are usually modeled using the RANS approach. In this approach, an instantaneous value of any variable is decomposed into a mean obtained by averaging over an appropriate time interval and the corresponding fluctuating component. Basic governing equations are formulated by appropriate averaging. A book by Ranade (2002) may be referred to for more details on the various approaches for formulating governing equations, including those for turbulent, multiphase, and reacting flows. Here we present the basic governing equations formulated using the RANS approach.

Typical numerical solution of the model equations is based on an iterative algorithm. Every iteration consists of solution of the overall continuity and momentum equations as the first step, which is followed by solution of governing equations for scalar quantities (species mass fractions, temperature, etc.). Governing equations for simulating PC fired boilers are therefore discussed in the same order here.

4.1.1 Overall Continuity and Momentum Balance Equations

The flow in a PC fired boiler is turbulent. The characteristic velocity scales are much smaller than the velocity of sound in the gas phase. Therefore, the flow can be represented without considering full compressibility effects. The gas-phase overall mass balance equation can be written as

$$\frac{\partial \rho}{\partial t} + \nabla \cdot (\rho U) = \sum_k S_k \tag{4.1}$$

where ρ is density, U is instantaneous velocity, and S_k is the source of species k for the gas phase from the particle phase. The first term on the left-hand side is the rate of change of mass with time; the second term on the left-hand side is the rate of change of mass in the x, y, and z directions; and the term on the right-hand side is the net source term for the gas phase from the solid phase. The gas-phase density ρ can be estimated using the ideal gas law (density as a function of molecular weight, pressure, and temperature). Considering low-velocity flows and significantly smaller variation in pressure compared to temperature, the gas-phase density can be estimated using the reference pressure instead of the local pressure.

In the RANS approach, an instantaneous value of any variable (ϕ^i) is decomposed into a mean obtained by averaging over an appropriate time interval and fluctuating component (ϕ') as

$$\phi^i = \bar{\phi} + \phi' \tag{4.2}$$

The overbar denotes time averaging. The Reynolds average form of the steady-state mass (overall) conservation equation for the gas phase can be written as

$$\nabla \cdot (\rho \bar{U}) = \sum_k \bar{S}_k \tag{4.3}$$

To simplify the equations, the overbars on quantities are omitted in the subsequent discussion. The variables, by default, indicate Reynolds averaged variables unless indicated otherwise. The formulation of S_k is discussed in the next section while discussing species conservation equations. Velocity appearing in the overall conservation equation is obtained by solving momentum equations.

The Reynolds averaging of the momentum balance equations leads to the appearance of new terms in the governing equations, which can be interpreted as the "apparent" stress gradient and additional sources associated with the turbulent motion. Hence, it becomes necessary to introduce a "turbulence model" that relates the new unknown terms to known terms in order to close the set of governing equations. The RANS-based turbulence models can be grouped into two classes: one that uses the concept of eddy or turbulent viscosity and another that does not. The latter class of models involves algebraic or differential forms of Reynolds stress models (RSMs). Considering the complexities of flow processes occurring in PC fired boilers, usually RANS models based on turbulent viscosity concepts are used. The Reynolds averaged momentum equations using the turbulent viscosity approach can be written as

$$\frac{\partial}{\partial t}(\rho U) + \nabla \cdot (\rho UU) = -\nabla p - \nabla \cdot \tau_{eff} + \rho g + F \qquad (4.4)$$

where p is the Reynolds averaged pressure, τ_{eff} is the effective shear stress tensor (viscous and Reynolds stresses), and F is the interphase momentum transfer (external force that arises from interaction with the dispersed phase).

The effective shear stress tensor can be written using the turbulent viscosity concept as

$$-\tau_{eff} = (\mu + \mu_T)\left(\frac{\partial U_i}{\partial x_j} + \frac{\partial U_j}{\partial x_i}\right) - \frac{2}{3}\delta_{ij}\left((\mu + \mu_T)\frac{\partial U_k}{\partial x_k} + \rho k\right) \qquad (4.5)$$

where δ_{ij} is the Kronecker delta function ($\delta_{ij} = 1$ if $i = j$, and $\delta_{ij} = 0$ if $i \neq j$); μ is the coefficient of viscosity; μ_T is the turbulent viscosity; and k is the turbulent kinetic energy. U_i is the mean velocity in the x_i direction.

The turbulent kinetic energy is defined as

$$k = \frac{1}{2}\overline{u_i u_i} \qquad (4.6)$$

The interphase momentum transfer term F can be obtained by simulating the trajectories of dispersed-phase particles (using the equation of motion for each dispersed-phase particle). It should be noted that depending on the degree of coupling (one-way, two-way, or four-way), the solutions of continuous and dispersed phases interact with each other. For one-way coupling, particle motion is influenced by the continuous phase, whereas the continuous phase is not influenced by presence of the dispersed phase. This is usually applicable for very dilute flows of inter-solids. For a PC fired boiler, the solid particles interact significantly with the gas phase and therefore influence the gas-phase flow field. However, because flows in a PC fired boiler are rather dilute, particle-particle interactions such as collision and coalescence/agglomerations (four-way coupling) can be neglected. Gas-solid flow is therefore modeled with the Eulerian–Lagrangian framework with two-way coupling. The two-way coupling between the continuous and dispersed phases occurs through interphase source terms such as F.

The momentum exchange from dispersed-phase particles to the continuous phase is just the opposite of the momentum transfer rate due to various forces exerted by the continuous phase on the dispersed-phase particles. These sources can be computed by summing the changes in momentum (or enthalpy or mass) of a particle as it passes through the control volume for the continuous phase, over all the particles passing through that control volume. This requires simulation of the motion of dispersed-phase particles. The motion or trajectories of dispersed-phase particles can be simulated by solving the momentum balance of the dispersed phase. The dispersed phase momentum balance on a single dispersed particle can be written as (Auton, 1983):

$$M_P \frac{dU_P}{dt} = F_p + F_D + F_G + F_{VM} + F_L + F_H + F_O \tag{4.7}$$

where M_p and U_p represent the mass and instantaneous velocity vector of the particle, respectively. The subscript p denotes the particle phase.

The right-hand side of Equation (4.7) is the net force acting on the particle: F_p is the force due to the continuous-phase pressure gradient, F_D is the drag force, F_G is the force due to gravity, F_{VM} is the virtual mass or added mass force (due to acceleration of the continuous phase), F_L is the lift force due to shear or vorticity of the continuous phase, F_H is the Basset history force due to the development of a boundary layer around the dispersed particle, and F_O is other forces, such as thermophoretic and Brownian forces.

The virtual mass force is not significant for gas-solid systems. The lift force, Basset history force, and other forces are usually much smaller than the drag, pressure, and gravity forces, and therefore are usually neglected while modeling flow processes in PC fired boilers. The sum of forces due to the continuous-phase pressure gradient F_p and due to gravity F_G can be written as

$$F_p + F_G = V_P \nabla p - \rho_P V_P g \tag{4.8}$$

where p is the pressure in the continuous phase, g is a gravitational constant, and ρ_P and V_P are the density and volume, respectively, of the particle. The drag force F_D can be written as

$$F_D = -\frac{\pi}{8} C_D \rho d_P^2 \left| U_P - U^i \right| \left(U_P - U^i \right) \tag{4.9}$$

where U^i denotes the instantaneous continuous-phase velocity, d_p is the particle diameter, and C_D is a drag coefficient. The drag coefficient can be estimated using a variety of correlations based on the applicable range of Reynolds number. Some of the widely used correlations for estimating the drag coefficient are listed in Appendix A4.2.1 and for multi-particle systems in Appendix A4.2.2 of Ranade (2002). Once the velocity flow field is calculated from the above force balance equations, the trajectory of any dispersed phase particle can be calculated as

$$\frac{dx}{dt} = U_P \qquad (4.10)$$

The interphase momentum transfer term F can be obtained using simulated particle trajectories and appropriate area and volume averaging. It is usually necessary to solve the governing equations of dispersed and continuous phases iteratively until one obtains the converged solution. More details on these can be found in Ranade (2002) and references cited therein.

It should be noted that the presence of turbulence will influence particle trajectories. The gas-phase momentum balance equation also involves an unknown term, that of turbulent viscosity. It is essential to use appropriate models to estimate effective or turbulent viscosity as well as the influence of turbulence on particle trajectories. By analogy with the kinetic theory of gases, turbulent viscosity can be related to the characteristic velocity and length scales of turbulence (u_T and l_T, respectively) as

$$\mu_T \propto \rho \, u_T \, l_T \qquad (4.11)$$

Several different turbulence models have been developed to devise suitable methods/equations to estimate these characteristic length and velocity scales to close the set of equations. Excellent reviews describing the relative merits and demerits of models pertaining to this class are available (see Ranade 2002 and references cited therein). For internal flows where characteristic length scales as well as velocity scales vary within the domain, it is recommended to use two-equation turbulence models. There are several different two-equation models proposed in the literature. All these models employ a modeled form of turbulent kinetic energy (to estimate the characteristic velocity scale). The choice of the second model transport equation, from which the length scale is determined, is the main differentiating factor among these models. The turbulent energy dissipation rate, ε, is one of the most widely used choices for the second transport variable of the two-equation turbulence models. In the standard k–ε model of turbulence, turbulent viscosity is related to k and ε by the following equation:

$$\mu_t = \frac{C_\mu \rho k^2}{\varepsilon} \qquad (4.12)$$

where C_μ is an empirical coefficient.

To close the set of equations, the local values of k and ε are obtained from the following transport equations of k and ε (Launder and Spalding, 1972):

$$\frac{\partial(\rho U_i k)}{\partial x_i} = \frac{\partial}{\partial x_i}\left(\frac{\mu_T}{\sigma_k}\frac{\partial k}{\partial x_i}\right) + G - \rho\varepsilon \tag{4.13}$$

$$\frac{\partial(\rho U_i \varepsilon)}{\partial x_i} = \frac{\partial}{\partial x_i}\left(\frac{\mu_T}{\sigma_\varepsilon}\frac{\partial \varepsilon}{\partial x_i}\right) + \frac{\varepsilon}{k}\left(C_1 G - C_2 \rho\varepsilon\right) \tag{4.14}$$

where G is the turbulence generation term and can be written as

$$G = \frac{1}{2}\mu_T\left[\nabla\overline{U} + \left(\nabla\overline{U}\right)^T\right]^2 \tag{4.15}$$

The model parameters appearing in these equations are not truly universal but are functions of the characteristic flow parameters.

One of the weaknesses of the standard k–ε model is that it over-predicts turbulence generation in regions where the mean flow is highly accelerated or de-accelerated. In a few cases, the standard k–ε model can predict very high values of the turbulent viscosity term. This is generally observed to start from the nose section (FEGT [furnace exit gas temperature] location) of a PC fired boiler and then proliferates until the outlet of the boiler. Several attempts have been made to enhance the applicability of the k–ε model by modifying either the empirical parameters or transport equations to suit the specific requirements of different types of flows. It will not be possible to discuss all the proposed modifications of the k–ε model here. There are two modifications—namely, the renormalized group version of the k–ε model (RNG k–ε model) and the realizable k–ε model—that are the most widely used versions in addition to the standard k–ε model.

The RNG version is formulated using RNG methods that represent the complex dynamics in terms of so-called "coarse-grained" equations governing the large-scale, long-time behavior. The term "realizable" means that the model satisfies certain mathematical constraints on the normal stresses, consistent with the physics of turbulent flows. This is achieved by making the parameter C_μ variable instead of a constant. Both versions also modify the transport equation of the turbulent energy dissipation rate (ε). The details of these equations are not discussed here for the sake of brevity; the interested reader can refer to specialized references on turbulence models for further information; see Ranade (2002) and references cited therein. The RNG form of the k–ε model uses the following equation to estimate the effective viscosity from a knowledge of k and ε:

$$\mu_{eff} = \mu\left(1 + \sqrt{\frac{C_\mu}{\mu}\frac{k}{\sqrt{\varepsilon}}}\right)^2 \tag{4.16}$$

TABLE 4.1

Brief Review of Turbulence Models used for PC Boiler Simulations

Turbulence Model	Ref.	Boiler Capacity (MWe)
Standard $k–\varepsilon$	Fang et al. (2012)	200 MW, tangentially fired
	Asotani et al. (2008)	40 MW, tangentially fired
	Diez et al. (2008)	600 MW, tangentially fired
	Belosevic et al. (2006)	210 MW, tangentially fired with burners mounted on the wall
	Kumar and Sahu (2007)	210 MW, tangentially fired
	Pallares et al. (2005)	139 MW, front wall fired
	Guo (2003)	Test combustor
	Yin et al. (2002)	600 MW, tangentially fired
RNG $k–\varepsilon$	Fan et al. (1999)	600 MW, tangentially fired
Standard and RNG $k–\varepsilon$	Hao et al. (2002)	600 MW, tangentially fired
	Fan et al. (2001)	600 MW, tangentially fired
	Fan et al. (1998)	300 MW, W-shaped boiler
Standard and Realizable $k–\varepsilon$	He et al. (2007)	Tangentially fired

This form allows extension to low Reynolds numbers and near-wall flows, unlike the standard $k–\varepsilon$ model, which is valid only for fully turbulent flows.

The standard $k–\varepsilon$ turbulence model and some of its variants, such as RNG or the realizable $k–\varepsilon$ turbulence model, have been widely used (Asotani et al., 2008; Belosevic et al., 2006; Pallares et al., 2005; Yin et al., 2002; Fan et al., 2001). A brief review of the turbulence models used for PC fired boiler simulations is provided in Table 4.1. Fan et al. (2000) have evaluated the influence of the turbulence models on predicted velocity profiles in PC fired boilers. These results are reproduced in Figure 4.1.

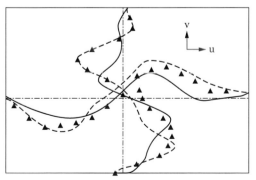

Symbols: Experimental data
—— : Standard k-ε model
– – – : RNG k-ε model

FIGURE 4.1
Comparison of velocity profiles simulated using standard and RNG $k–\varepsilon$ model with experimental data. Reprinted from *Chemical Engineering Journal*, *81* (1-3), Fan, J., Qian, L., Ma, Y., Sun, p. and Cen, K., Computational modelling of pulverized coal combustion process in tangentially fired furnaces, 261–269 (2001) with permission from Elsevier.

It was observed that the predictions of RNG k–ε were more accurate than the standard k–ε model. However, significant experimental data from PC fired boilers are not available to really discriminate among the various versions of the k–ε model. The turbulence flow field estimated by any version of the model can then be used to simulate the influence of turbulence on particle trajectories.

When the continuous-phase flow field is turbulent, then in order to appropriately calculate interphase exchange sources, a sufficiently large number of particle trajectories must be simulated. In addition to a large number of trajectories, there are many factors that are important to account for turbulent gas-solid interactions (Shirolkar et al., 1996), such as

- Size of the particle and characteristic length scale of the eddy
- Density difference between gas and solid
- Gas velocity fluctuation around the particle
- Particle integral time scale: time scale for which the particle maintains its original velocity before it undergoes a change in velocity due to the effect of turbulence
- Particle relaxation time: rate of the response of the particle to the relative velocity between fluid and the particle

It should be noted that the instantaneous velocity of the gas phase must be known for solving the momentum balance of individual particles. The instantaneous velocity can be estimated as

$$U^i = U + u \qquad (4.17)$$

where u is a fluctuating velocity component and U is the Reynolds averaged velocity. Several models have been proposed to estimate the instantaneous velocity; see, for example, reviews by Sommerfeld (1993) and Gouesbet and Berlemont (1999). Two of the commonly used models to estimate instantaneous fluid velocity from the time-averaged flow field of the continuous phase are the discrete random walk (DRW) model (Gosman and Ioannides, 1981) and the continuous random walk (CRW) model (Thomson, 1987). These models differ in the way in which the instantaneous fluid velocity is estimated from a knowledge of gas-phase turbulence.

In the DRW model, which is also called the "eddy lifetime model," the fluctuating velocity is added to the mean gas velocity for a constant time interval given by the characteristic lifetime of gas phase eddies. In this case, each eddy was characterized by the

- Gaussian distributed random velocity fluctuation u
- Eddy lifetime τ_e

The values of u that prevail during the lifetime of the turbulent eddy were assumed to obey a Gaussian distribution during the lifetime of an eddy and were estimated as

$$u = \zeta \sqrt{2k/3} \qquad (4.18)$$

Here, ζ is a normally distributed random number and k is the turbulent kinetic energy. The characteristic eddy lifetime can be defined either as a constant

$$\tau_e = 2\,T_L \qquad (4.19)$$

or as a random variation about T_L:

$$\tau_e = -\,T_L \log(r) \qquad (4.20)$$

where r is a random number from 0 to 1.0. T_L is the integral time scale of turbulence and can be estimated from knowledge of the turbulent kinetic energy (k) and turbulent kinetic energy dissipation rate (ε) as (Michaelides, 1997)

$$T_L = 0.15 \frac{k}{\varepsilon} \qquad (4.21)$$

The solid particle is assumed to interact with the fluid-phase eddy over this eddy lifetime. When the eddy lifetime is reached, a new value of the instantaneous velocity is obtained by applying a new value of ζ in Equation (4.18). The DRW approach leads to reasonable results and is widely used for a number of applications (see, for example, Chen and Crowe, 1984; Sommerfeld, 1990).

In the CRW model, the instantaneous fluid velocity is obtained by solving the Langevin equation (Thomson, 1987), which may provide a more realistic description of turbulent eddies, but at the expense of greater computational effort. These models correlate the velocity fluctuation experienced by a particle at the new location with the fluctuation at the previous location using an exponentially decaying correlation function (the Langevin velocity correlation function). In inhomogeneous turbulence, however, it was found that the Langevin model fails and may lead to the accumulation of fluid particles in the region of low turbulence (Sommerfeld, 1993). Berlemont et al. (1990) have proposed a more sophisticated approach, the so-called "correlation slaved approach," to estimate instantaneous velocity. They considered a correlation matrix that evolves along the trajectory (also see Gouesbet and Berlemont, 1999). Their model, however, needs much larger computational resources than the above two approaches. Most PC fired boiler simulations typically use either DRW- or CRW-like models.

Once the instantaneous velocity is obtained, particle trajectories can be simulated. To introduce two-way coupling, it is necessary to calculate the source terms in the balance equations of mass, momentum, and energy for the continuous phase. With such source terms, the continuous-phase flow field needs to be solved again, which is later used to calculate new trajectories. For strongly coupled flows such as in PC fired boilers, convergence may be quite difficult to reach, and appropriate care and special precautions should be taken to steer simulations toward convergence (see Kohnen et al., 1994). More details and further discussion on these topics can be found in Ranade (2002), Gouesbet (1998), and Shirolkar (1996). In the Eulerian–Lagrangian approach, dispersed-phase particle size distributions can be conveniently accommo-dated. In large complex systems such as PC fired boilers, it is usually not pos-sible to track all the injected particles. Representative particles are therefore tracked, with each tracked particle representing multiple particles.

While simulating trajectories of dispersed-phase particles, appropriate boundary conditions must be specified. Inlet or outlet boundary conditions require no special attention. At impermeable walls, however, it will be nec-essary to represent collisions between particles and the wall. A particle can reflect from the wall via an elastic or inelastic collision. Suitable coefficients of restitution representing the fraction of momentum retained by the particle after the collision must be specified at all wall boundaries. In some cases, particles may stick to the wall or may remain very close to the wall once they collide with the wall. Special boundary conditions should be developed to model these situations.

4.1.2 Species Balance Equations

The steady-state species conservation equation can be written as

$$\nabla \cdot (\rho U m_k) = \nabla \cdot \left(\left(\rho D_{km} + \frac{\mu_t}{Sc_t} \right) \nabla m_k \right) + R_k + S_k \qquad (4.22)$$

The term on the left-hand side is the rate of change of mass of component k. The first term on the right-hand side is the rate of change of the mass of component k due to molecular and turbulent diffusion; the second term is the rate of consumption or generation of component k due to gas-phase reac-tions; and the third term is the rate of consumption or generation of compo-nent k due to the solid phase.

In Equation (4.22), m_k is mass fraction of species k, D_{km} is the diffusion coefficient for species k in the mixture, Sc_t is the turbulent Schmidt number ($Sc_t = \frac{\mu_t}{\rho D_t}$ where is the turbulent viscosity and D_t is the turbulent mass dif-fusivity; a typical value of Sc_t is 0.7); R_k is the net rate of production of species

k by all gas-phase chemical reactions; and S_k is the source of species k transferred from the dispersed phase to the continuous phase.

The turbulent Schmidt number measures the relative diffusion of momentum and mass due to turbulence and is on the order of unity in all turbulent flows. The majority of CFD modeling studies have used $Sc_t = 0.7$, as recommended by Spalding (1971). On the other hand, Launder (1978) pointed out that Sc_t shows a value of 0.9 for turbulence near the wall. A few researchers have used $Sc_t = 1.3$ for simulating PC fired boilers (see, for example, Guo, 2000; Fan et al., 1999, 1998). The formulation of R_k and S_k follows.

The volatile matter released from coal reacts with oxygen present in the gas phase. A typical volatile material has many components, such as light gases, tar, CH_4, etc. There are sophisticated models such as CPD (chemical percolation devolatilization), FG-DVC (functional group depolymerization, vaporization, cross-linking), and FLASHCHAIN for estimating the yield and composition of volatile products released from coal (refer to Chapter 3, Table 3.1). These models, however, often require large input parameters that are seldom available. In the absence of such data, the volatile can be represented by $C_xH_yO_z$, and a stoichiometric balance for it can be obtained for oxidation of the volatile material. The values of x, y, and z can be obtained from the proximate and ultimate analyses of coal. To illustrate this approach, a specific Indian coal was characterized, and the volatile material from this coal was represented by a single species as $CH_{2.08}O_{0.38}$. The following lumped gas-phase reactions were assumed:

$$CO_{(g)} + 0.5\,O_{2(g)} \rightarrow CO_{2(g)} \quad \Delta H^o = -283\ kJ/mol$$

$$CH_{2.08}O_{0.38\,(g)} + 1.33\,O_{2\,(g)} \rightarrow CO_{2(g)} + 1.04\,H_2O_{(g)} \quad \Delta H^o = -271\ kJ/mol$$

Carbon monoxide (CO) is released from the char oxidation reaction (a heterogeneous reaction). The generated CO is diffused out of the pores of the char-ash matrix and then released into the surrounding bulk gases where it reacts with the oxygen to produce CO_2. The source of species k due to heterogeneous reactions, S_k, is discussed after discussing the formulation of rates of homogeneous reactions (R_k) of species k.

The homogenous gas-phase reactions are treated using a species transport approach with either the kinetics or mixing controlled reaction rates (Li et al., 2003; Zhou et al., 2002; Fan et al., 2001). Alternatively, these can also be modeled using a single or two-mixture fraction approach (He et al., 2007; Pallares et al., 2005; Yin et al., 2002). A species transport and finite chemistry-based model solves $(n-1)$ conservation equations for n species in the defined problem and accounts for the diffusion, interphase mass transfer, and generation/destruction of species in the gas phase. The rate of gas-phase combustion at low temperature is kinetic rate controlled. Once the flame is ignited and the temperature is high, the reaction rate may become mixing limited. This can be captured by an Arrhenius-type kinetic rate/eddy dissipation model. The gas-phase reaction rate is evaluated based on the minimum of the reaction rate estimated by the Arrhenius-type kinetic rate model and the eddy dissipation model. The eddy dissipation model is based on the work of Magnussen and Hjertager (1976) and describes the turbulence–chemistry interaction. The net rate of production of species k due to reaction r, $R_{k,r}$, is given by the smaller (i.e., limiting value) of the two expressions below:

$$R_{k,r} = v'_{k,r} M_{w,k} A \rho \frac{\varepsilon}{k} \min_R \left(\frac{Y_R}{v'_{R,r} M_{w,R}} \right) \tag{4.23}$$

$$R_{k,r} = v'_{ki,r} M_{w,k} A B \rho \frac{\varepsilon}{k} \left(\frac{\sum_P Y_P}{\sum_j^N v''_{j,r} M_{w,j}} \right) \tag{4.24}$$

where Y_p is the mass fraction of any product species p, Y_R is the mass fraction of a particular reactant R, A is an empirical constant = 4.0, and B is an empirical constant = 0.5. In these equations, the chemical reaction rate is governed by the large-eddy mixing time scale, k/ε, as in the eddy-breakup model of Spalding (1970).

The molar rate of creation/destruction of species k in reaction r in Arrhenius form can be written as

$$R_{k,r} = \left(v'_{k,r} - v_{k,r} \right) K_r \prod_l \left[C_{l,r} \right]^{\eta_{l,r}} \tag{4.25}$$

where $C_{l,r}$ is the molar concentration of each reactant l^{th} species in reaction r, $\eta'_{l,r}$ is the exponent for each lth reactant in reaction r, $v'_{k,r}$ and $v_{k,r}$ are stoichiometric coefficients for k^{th} species as product and reactant, respectively, and K_r is the kinetic rate constant for any reaction r and can be expressed as

$$K_r = A_r \, e^{(-E_r/RT)} \tag{4.26}$$

Here, A_r and E_r are the preexponential factor and activation energy for the gas-phase reaction r, T is the gas temperature, and R is the universal gas constant. The gas-phase reaction rate is evaluated based on the minimum rate obtained from the above three equations (Equations (4.23) to (4.25)).

Another approach to handle this complex chemistry is mixture fraction based, in which a single conserved scalar (mixture fraction) is solved instead of transport equations for individual species. The reacting system is treated using either chemical equilibrium calculations or by assuming infinitely fast reactions (mixed is reacted approach). In this approach, transport equations are formulated for the time-averaged mixture fraction and time-averaged variance of the mixture fraction. Based on these two values, individual mass fractions of fuel and oxidants (see Toor, 1975; Jones and Whitelaw, 1982) can be obtained. To calculate the time-averaged values of species concentrations, the probability density function (PDF) approach is used. Several mathematical functions have been proposed to express the PDF. The presumed PDF methods assume a form for the PDF rather than computing it. Such methods also have been extended to nonadiabatic flow processes such as coal combustion. In nonadiabatic processes, the local thermochemical state is also related to enthalpy, and then it is necessary to employ a joint PDF for mixture fraction and enthalpy. Most of the literature on simulating PC fired combustion has adopted or presumed a PDF approach. Case studies on the PDF-based mixture model can be found in Vuthaluru and Vuthaluru (2006), Sheng et al. (2004), and Yin et al. (2002). Here we use the species transport approach while illustrating application of the CFD model to simulate a typical 210-MWe PC fired boiler. This illustrative application and the results are discussed in the next section.

The source of species k due to the dispersed phase is calculated by solving species conservation equations for the particle phase. Species conservation equations for the discrete phase can be written as

$$\frac{dM_k}{dt} = S_{pk} \tag{4.27}$$

where M_k is the mass of species k present in the particle. The particle source term of species k, S_{pk} can be formulated by considering various particle-level phenomena of interest, such as devolatilization and surface reaction-char combustion. Hence, S_{pk} can be written as

$$S_{pk} = \left[\frac{dM_v}{dt} + \frac{dM_c}{dt} \right] \tag{4.28}$$

where, M_v and M_c are the mass of volatiles and char, respectively, in a coal particle.

The combustion of pulverized coal has two distinct processes occurring after entering the furnace: devolatilization and char oxidation. Devolatilization of coal in a PC fired boiler is quite fast and accomplished within a fraction of a second. Char oxidation is a comparatively slower step. The source terms were formulated using various types of devolatilization and combustion models discussed in Chapter 3. For coal devolatilization, a single-step reaction-based simple model provided by Badzioch and Hawksley (1970) is widely adopted as

$$\frac{dM_v}{dt} = -A_v\, e^{(-Ev/RT_P)}\, M_v \tag{4.29}$$

where M_v is the mass of volatile present in coal particle at any time, A_v is the preexponential factor, E_v is the activation energy for devolatilization, and T_p is the temperature of the particle. The devolatilization kinetic parameters are typically obtained from TGA (thermogravimetric analysis) experiments.

Char combustion is commonly assumed as char is oxidized to CO via the following reaction:

$$C_{(s)} + 0.5\, O_{2\,(g)} \rightarrow CO_{\,(g)} \qquad \Delta H° = -110\ kJ\,/\,mol$$

The rate for this reaction is typically expressed as

$$Rate = K_c A_p P_{o2}^n \tag{4.30}$$

where the kinetic rate constant (K_c) for the char oxidation reaction is represented as

$$K_c = A_c\, e^{(-E_c/RT_P)} \tag{4.31}$$

where A_c is the preexponential factor and E_c is the activation energy for char combustion. A_p is the area of the particle, P_{O2} is the partial pressure of oxygen, and n is the order of reaction with respect to oxygen. The kinetic parameters for char combustion are usually obtained via DTF (drop-tube furnace) (see Chapter 3 for more details). Char combustion is often influenced by gas-solid mass transfer. The rate of char oxidation based on the apparent kinetic parameters for any particle can be written as (Baum and Street, 1970; Field, 1969)

$$\frac{dM_c}{dt} = -A_p\, \frac{K_c K_d}{K_c + K_d}\, P_{O_2} \tag{4.32}$$

where M_c is the mass of char present in coal at any time. The bulk gas-phase diffusion coefficient for the oxidant (Field, 1969) can be given as

$$K_d = \frac{5 \times 10^{-12}}{d_p} \left(\frac{T_g + T_p}{2} \right)^{0.75} \tag{4.33}$$

Different approximations about how particle size and particle density vary during devolatilization and combustion processes have been reported (see, for example, Smith, 1971). Here, for the sake of simplicity, the particle size was assumed to remain unchanged during devolatilization as well as during combustion processes. The particle density was varied to account for changes in particle mass during these processes. The methodology presented here, however, can be extended in a straightforward manner to incorporate more complex approximations.

The overall combustion process includes moisture vaporization, devolatilization, char oxidation, and homogeneous gas-phase reactions. Various models available to calculate volatile yield include the constant rate model (Baum and Street, 1970; Pillai, 1981), the single-step (Badzioch and Hawksley, 1970) and two-step Arrhenius kinetic rate models (Kobayashi et al., 1977), the CPD model (Fletcher et al., 1992), and more. It has been recognized that single-step models can successfully predict the devolatilization of coal, provided that the appropriate coal-specific kinetic parameters are known (Jones et al., 1999; Brewster et al., 1995). The char oxidation has been computed mainly using the global/apparent kinetic rate equation (Baum and Street, 1970) that couples the effect of internal ash layer diffusion with intrinsic reactivity. Recently, more advanced combustion models, such as the char burnout kinetic (CBK) model, that can predict the effects of heterogeneity in maceral content, thermal deactivation, and ash inhibition in the late stages of combustion have been used for predicting unburned char in ash (Pallares et al., 2005; Hurt et al., 1998). Global/apparent kinetic models are particularly useful where coal-related information such as porosity, pore diameter, the effectiveness factor, surface area of char, etc. is not readily available. Several efforts have been made to improve the prediction capabilities of the present CFD combustion models for obtaining quantitative predictions of char burnout (Pallares et al., 2005; Backreedy et al., 2005; Hurt et al., 1998).

By simulating trajectories and variation of mass fractions and mass of injected particles along the trajectories, corresponding sources of the gas phase can be formulated. Formulation of continuous-phase sources from particle-level sources is discussed by Ranade (2002).

4.1.3 Energy Balance Equations

The energy balance equation for the gas phase can be written as

$$\nabla \cdot (\rho U h) \;=\; \nabla \cdot (k_{eff} \nabla T) - \nabla \cdot \left(\sum_k h_k j_k \right) + S_h \tag{4.34}$$

where h is an enthalpy and k_{eff} is the effective thermal conductivity of the gas. The term on the left-hand side represents the change in enthalpy due to convection. The terms on the right-hand side represent changes in enthalpy due to conduction, diffusive mass fluxes (j_k), and volumetric sources of enthalpy (due to, say, chemical reactions and/or the dispersed phase). The changes in enthalpy due to pressure and viscous dissipation are neglected in this balance equation because they are very small in the case of PC fired boilers.

The volumetric source term S_h is the sum of the heat of chemical reactions ($S_{h,rxn}$) occurring in the gas-phase, source term because of the discrete phase (S_Q), radiation (S_R), and heat sink to water walls and heat exchangers (S_E), and may be written as

$$S_h = S_{h,rxn} + S_Q + S_R - S_E \tag{4.35}$$

The heat added from the discrete phase is:

$$S_Q = \sum_j \frac{\left[(1-f_{heat})\left(\frac{dM_c}{dt}\cdot N_p\right)H_c + Q_{rad} + Q_{conv}\right]_j}{V} \tag{4.36}$$

The f_{heat} is the fraction of heat absorbed by the particle, M_c is the mass of carbon present in coal, N_p is number of paticles (refer Equation (3.43)), H_c is the heat released during char oxidation, and Q_{rad} and Q_{conv} are the radiative and convective heat transfers between gas and particle, respectively.

Thermal radiation in the furnace of a boiler is the dominant mode of heat transfer. The flue gas composition in the furnace (particularly the CO_2 and H_2O concentrations) affects the absorption coefficient, emissivity, and therefore the radiative heat transfer. The underlying assumptions about the gray or non-gray gas, specular or diffusive surfaces, and isotropic or anisotropic scattering play an important role in effective heat transfer. The contribution of absorbing and emitting participating media (such as coal, char, ash, and soot particles) is also important in determining overall radiative exchange. The variation in surface properties of the tube walls due to particle deposition and aging also affects the radiative heat transfer. Therefore, capturing radiative heat transfer in a PC fired boiler is one of the most complex problems due to the participation of non-gray gases, the presence of particulates, and the three-dimensional (3D) complex geometry. Estimation of radiative heat transfer has two basic parts:

1. Solution to radiative heat transfer equation (RTE)
2. Estimation of the radiative properties of the gases and particles

The accuracy of the radiation model depends on how accurately the above two aspects are captured. If the gas phase in a PC fired boiler is assumed as non-gray, then the intensity of radiation within the medium becomes a strong function of the spectral variation, and integration of the RTE becomes too complex. This requires that the RTE be solved multiple times for each wavelength, thus leading to large demands on computational resources. Therefore, more often than not, large combustion systems such as PC fired boilers use the gray gas assumption; that is, the emissivity of the gas does *not* depend on wavelength. A large body of literature exists on various aspects of radiative heat transfer (see, for example, Modest (2003), Howell and Siegel (2002), Viskanta (1987), Hottel and Sarofim (1967), Coelho (2007)). Eaton et al. (1999) have reviewed radiative heat transfer, particularly in combustion systems. It will be useful to consult these papers and reference cited therein. Here we briefly discuss and present some governing equations.

The RTE in a participating gray medium is the change in intensity of thermal radiation at position r along the directions ω and for steady state (as the speed of light is several magnitudes faster than the underlying time scale of the process), it can be written as

$$\frac{dI(r,\omega)}{ds} = -(a+\sigma)I(r,\omega) + aI_b + \frac{\sigma}{4\pi}\int_{4\pi} I(r,\omega)\Phi\, d\omega \qquad (4.37)$$

where $I(r,\omega)$ is the intensity of radiation at position r and direction ω, a is the overall absorption coefficient, σ is the scattering coefficient, and Φ is the scattering phase function.

The first term on the right-hand side of Equation (4.37) represents the intensity of radiation loss due to absorption and scattering in the path length ds of the medium containing gas and particles. The second terms represent emissions added to the radiation intensity from the path length ds, and the third term represents radiation intensity added from internal scattering.

Equation (4.37) is integrated to provide the radiative source term (Q_r) in the energy balance equation as

$$Q_r = a_g\left(\int_{4\pi} I\, d\omega - 4E_b\right) \qquad (4.38)$$

where E_b is the blackbody emissive power of the gas.

For the gray approximation, the blackbody emissive power is expressed as

$$E_b = \sigma T_g^4 \qquad (4.39)$$

where σ is the Stefan Boltzmann constant = 5.67×10^{-8} W/m²-K⁴, and T_g is the local gas temperature.

Exact solutions of the RTE for practical problems such as radiative heat transfer in a boiler are not available (Modest, 2003; Howell and Siegel, 2002). Following are the main approaches to solve RTE:

- *Monte Carlo (MC) or statistical method* (Howell and Siegel, 2002): This method tracks the history of a large number of photons generated from a volume or surface as they are transported in the solution domain until the intensity of the photons reduces to a pre-described level. It is one of the most accurate and rigorous methods to simulate radiation. However, it is computationally very expensive due to tracking a very large number of photons.
- *Hottel zone method* (Hottel and Sarofim, 1967): The computation domain is divided into a number of zones having uniform temperature and radiation properties. The radiation heat exchange between each zone with all other zones is estimated based on the direct exchange areas. This approach is described in Chapter 5 and generally is more suitable for phenomenological-type models having a small number of zones/volumes.
- *Flux methods:* These methods assume that the intensity of radiation is uniform over defined intervals of the solid angle. This simplification reduces the current RTE (integro-differential) form to ordinary differential equations that can be easily solved using the current CFD framework in an efficient manner. The discrete ordinate (DO) method is one example.

The commonly used radiation approaches for boiler or combustion furnaces are listed in Table 4.2.

Approximate radiation models such as the P-1 harmonic series and DO models have been widely adopted in CFD codes for solving radiative heat transfer equations for specific application to PC fired boilers. These models have sufficient generality and conformance with control-volume formulations and can account for particulate effects (particle scattering). These approximate methods are preferred over more accurate methods such as MC methods because approximate methods are computationally quite efficient in estimating the radiative source term in reaction-turbulence coupled processes that are solved iteratively. P-1 is considered more suitable for combustion systems having a thick optical thickness (that is, $a \times L > 3$, where a is the absorption coefficient and L is width or depth of the furnace). The DO model is also used, and further information on the DO model can be obtained from Yin (2002) and Zhou et al. (2002) for its application to PC fired boilers. Here we present a brief discussion of the P-1 and DO radiation models for their application to PC fired boilers.

TABLE 4.2

Brief Review of Radiation Models Used for PC Fired Boiler Simulations

Radiation Model	Refs.	Boiler Capacity (MWe)
P-1	Asotani et al. (2007)	40 MW tangentially fired
	Vuthaluru et al. (2006)	500 MW, wall fired
	Backreedy et al. (2006)	1 MWth burner testing
Discrete ordinate (DO) model	He et al. (2007)	Capacity not available, tangentially fired
	Yin et al. (2002)	609 MW, tangentially fired
	Zhou et al. (2002)	600 MW, tangentially fired
Discrete transfer model	Diez et al. (2007)	600 MW, tangentially fired
	Xu et al. (2001)	300 MW, wall fired
	Lee and Lockwood (1981)	Method demonstrated for general combustor
Monte Carlo model	Fan et al. (2001)	300 MW, wall fired
	Hao et al. (2002)	600 MW, tangentially fired
	Marakis et al. (2000)	Cylindrical furnace; compared P-1 and MC and found predictions comparable
Six-flux model	Belosevic et al. (2006)	210 MW, tangentially fired

4.1.3.1 P-1 Radiation Model

The P-N model (Modest, 2003) is based on the expansion of the local radiation intensity in terms of orthogonal series of spherical harmonics, with truncation to N terms in the series and substitution into the moments of the differential form of the RTE. In the present study, the simplest and lowest-order P-1 approximation is discussed, which is adequate for most PC fired boiler applications.

The radiation transport equation for the P-1 model can be written as

$$\nabla \cdot (\Gamma \nabla G) = (a + a_p)G - 4\pi \left(a\frac{\sigma T^4}{\pi} + E_p \right) \qquad (4.40)$$

The quantity Γ is $\Gamma = \dfrac{1}{3(a + a_p + \sigma_p)}$ (4.41)

where G is the incident radiation and equal to $4\sigma T^4$, σ is the Stefan–Boltzmann constant, a is the absorption coefficient of the gas phase, and a_p is the equivalent absorption coefficient due to the presence of particulates and is defined as

$$a_p = \lim_{V \to 0} \sum_{n=1}^{N} \varepsilon_{pn} \frac{A_{pn}}{V} \qquad (4.42)$$

The equivalent emission E_p is defined as

$$E_p = \lim_{V \to 0} \sum_{n=1}^{N} \varepsilon_{pn} A_{pn} \frac{\sigma T_{pn}^4}{\pi V} \tag{4.43}$$

where ε_{pn}, A_{pn}, and T_{pn} are the emissivity, projected area, and temperature of the particle n, respectively. The summation is taken for all particles present in the volume V.

The equivalent particle scattering factor σ_p is given as

$$\sigma_p = \lim_{V \to 0} \sum_{n=1}^{N} \left(1 - f_{pn}\right)\left(1 - \varepsilon_{pn}\right) \frac{A_{pn}}{V} \tag{4.44}$$

where f_{pn} is the scattering factor associated with the nth particle. The equivalent particle scattering factor σ_p is computed during particle tracking.

The expression for $\nabla \cdot (\Gamma \nabla G)$ can be directly substituted into the energy equation to account for heat sources (or sinks) due to radiation.

4.1.3.2 Discrete Ordinate (DO) Radiation Model

The DO radiation model (Modest, 2003) solves the RTE for a finite number of discrete solid angles, each associated with a vector direction $\vec{s}$ fixed in the global Cartesian system (x, y, and z). The DO model transforms the transport equation for radiation intensity in the spatial coordinates (x, y, and z). The DO model solves as many transport equations as there are directions $\vec{s}$. The solution method is identical to that used for the fluid flow and energy equations. The DO model considers the RTE in the direction $\vec{s}$ as a field equation. Thus, Equation (4.37) is written as

$$\nabla \cdot \left(I(\vec{r},\vec{s})\vec{s}\right) + \left(a + \sigma_s\right) I(\vec{r},\vec{s}) = a I_b + \frac{\sigma}{4\pi} \int_{\Omega=0}^{4\pi} I(\vec{r},\vec{s}')\varphi\left(\vec{s},\vec{s}'\right) d\Omega' \tag{4.45}$$

The DO model also allows the non-gray gas approach, and Equation (4.45) can be written for each wavelength (λ) of radiation.

If q_{in} is the total radiant intensity on the walls, then some part of it is reflected diffusely and specularly, depending on the diffuse fraction f_d of the wall. Some of the incident radiation is absorbed at the surface of the wall, and some radiation is emitted from the wall surface. This can be broadly grouped as

- Emission from the wall surface $= \varepsilon_w \sigma T_w^4$
- Diffusely reflected energy $= f_d(1 - \varepsilon_w)q_{in}$

- Specularly reflected energy $= (1 - f_d)q_{in}$
- Absorption at the wall surface $= f_d \varepsilon_w q_{in}$

where f_d is the diffuse fraction. For a purely diffused wall, f_d is equal to 1 and there is no specularly reflected energy. For a purely specular wall, f_d is equal to 0 and there is no diffusely reflected energy. A diffuse fraction between 0 and 1 will result in partially diffuse and partially reflected energy.

There are four approximate approaches proposed for predicting the radiative properties of gases (Eaton et al., 1999; Modest, 2003; Viskanta, 1987):

1. *Line-by-line approach:* It depends on a detailed knowledge of every single spectral line, and an absorption coefficient is calculated for each single line. Such information is obtained from the HITRAN96 database, which is experimentally generated at atmospheric conditions, and the HITEMP database, which is an extrapolation of the HITRAN96 database to high temperature. As all combustion processes are at high temperature and HITEMP is an extrapolated database, its accuracy is not proven. Also, computation based on this method will require very large computations resources.

2. *Narrow band approach:* This approach gives an average absorption coefficient and intensity over a narrow spectral range, instead of the actual individual spectral line.

3. *Wide band approach:* This approach integrates the narrow band results across an entire band.

4. *Global approach:* Look-up tables for the properties are created based on the above three approaches. Based on the composition of the gases, and the operating pressure and temperature, the desired property is derived based on assigning a weighting factor to each gas species. The weighted sum of gray gas model (WSGGM) and its further extensions are the commonly adopted for engineering purposes and are based on the global approach.

Some numerical experiments were performed to understand the difference in the predictions of the P-1 and DO models. Instead of solving combustion, a simple numerical experiment of passing hot flue gas through the boiler was performed. The simulated values of overall heat transferred to the water wall in the furnace section, superheater, and the temperature profiles obtained using the P-1 and DO models were compared. A sample of this comparison is shown in Figure 4.2.

These numerical experiments were performed in a typical geometry of a 210-MWe PC fired boiler with a gas absorption coefficient of 0.5 and a scattering coefficient of 0. The wall emissivity was set to 0.6, and the wall temperature was set to 600K. It can be seen that the simulated values of overall

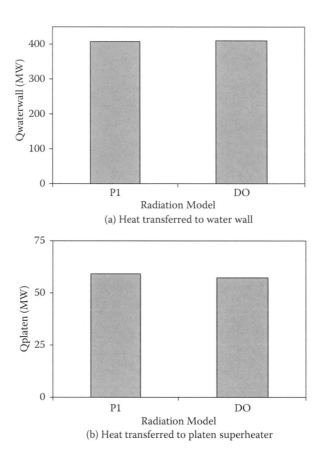

FIGURE 4.2
Comparison of P-1 and DO radiation models: (a) heat transferred to water wall and (b) heat transferred to platen superheater.

energy absorbed by the water walls or other internal heat exchangers are not significantly affected by the choice of radiation model. Therefore, either of these two models can be used to simulate radiative heat transfer in boilers.

The radiation model has parameters such as the absorption coefficient, scattering coefficient, and emissivity of the heat exchangers, and the sensitivity of such parameters can be also studied to understand their relative importance. The numerical experiment described above was therefore extended to investigate the sensitivity of the simulated results with other relevant parameters of radiative transport. The absorption coefficient of gas is a key parameter. The intensity of the incident radiation is attenuated because of absorption by the gas through which the radiation travels. The absorption leads to an exponential decay of incident radiation.

Therefore, the transmissivity of the homogeneous isothermal gas layer can be written as (Siegel et al., 1972)

$$\tau_i = e^{-a_i S} \tag{4.46}$$

where S is the thickness of the gas layer, and the proportionality constant a_i is known as the absorption coefficient and has units of reciprocal of length (1/m). The absorption coefficient, in general, is a function of the temperature, pressure, composition of material, and wavelength of incident radiation, which can be estimated (Modest, 2003) by the weighted sum of gray gas model (WSGGM). The WSGGM allows calculation of the local value of the absorption coefficient as a function of local mass fractions of water vapor and carbon dioxide (or of any other species). The simulated values of heat transferred to the furnace water wall and platen superheater for different values of absorption coefficients (0, 0.5, and 1 as constant values) and WSGGM based (composition, temperature, pressure dependent) are shown in Figure 4.3. All simulations were performed with a wall emissivity of 0.6 and a temperature of 600K.

When $a = 0$ is assigned to the absorption coefficient, the medium becomes completely transparent to the incident radiative energy due to which the gas shows low absorption and emittance. This leads to low heat being transferred to the water wall in the furnace zone and more heat is transferred to the platen superheater. The results show that the heat transferred to the water wall was about 250 MW when $a = 0$. When $a = 0.5$, 1, and WSGGM, the simulations predict that the heat transferred to the water wall will be 415, 380, and 355 MW, respectively. Variation of the local absorption coefficient predicted using the WSGGM is shown in Figure 4.4. Considering the wide variation in local composition and temperature in the PC fired boiler, the WSGGM model is recommended and was used in subsequent simulations of coal combustion in a boiler.

When a photon (or an electromagnetic wave) interacts with a gas molecule, it may be scattered, which results in a change in the direction of travel of the photon. It can be defined as the inverse of the mean free path traveled before the photon undergoes scattering. The influence of the scattering coefficient (σ_s) on predicted heat transfer is shown in Figure 4.5. It can be seen that with the P-1 radiation model, the value of the scattering coefficient has a marginal influence on the heat transferred to water walls and other internal heat exchangers such as platen superheaters. The emissivity of the wall was set to 0.6, and the absorption coefficient was set to 0.5 in these simulations. The scattering coefficient was therefore set to zero for further simulations.

Emissivity is a measure of a material's ability to absorb and radiate energy. The emissivity of water wall tubes decides the heat that will be absorbed in the furnace zone. Ash deposition is one major factor that affects the emissivity of water walls. The influence of the variation in emissivity on heat

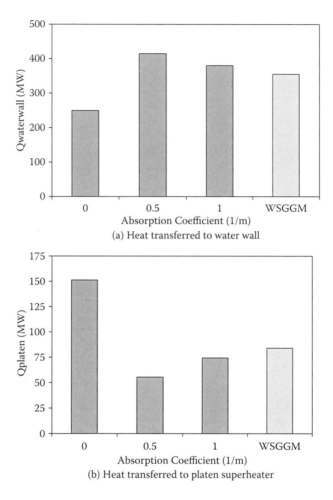

FIGURE 4.3
Influence of absorption coefficient on heat transfer in a typical 210-MWe PC fired boiler:
(a) heat transferred to water wall and (b) heat transferred to platen superheater.

transfer on the water wall and platen superheater is illustrated Figure 4.6
using the numerical experiment described previously.

The emissivity of the water wall adopted for boiler simulations ranges from
0.6 to 0.85. Asotani et al. (2008) have used 0.6 as the water wall emissivity for
a 40-MWe boiler, and 0.85 was used by Diez et al. (2008) and Backreedy et al.
(2006) for a 600-MWe boiler. It should be noted that, typically, the actual geometric details of the water walls such as shape, size, and pitch of the tube are
usually not accounted for in the CFD model (to simplify the grid generation);
the emissivity of water walls is treated as an adjustable parameter that can
be tuned to obtain the right FEGT, and the heat transferred in the furnace
section to the water wall.

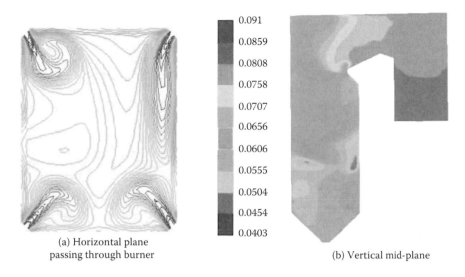

0.091	
0.0859	
0.0808	
0.0758	
0.0707	
0.0656	
0.0606	
0.0555	
0.0504	
0.0454	
0.0403	

(a) Horizontal plane
passing through burner

(b) Vertical mid-plane

FIGURE 4.4 (See Color Insert.)
Simulated distribution of absorption coefficient (1/m) using WSGGM: (a) horizontal plane passing through burner and (b) a vertical mid-plane.

Thermal radiation in the furnace is primarily modeled by the P-1 model (Asotani et al., 2008; Vuthaluru et al., 2006; Backreedy et al., 2006) or the DO model (He et al., 2007; Yin et al., 2002; Zhou et al., 2002). Other approaches, such as the six-flux model (Belosevic et al., 2006), Monte Carlo model (Fan et al., 2001; Howell, 1968), and discrete transfer model (Xu et al., 2001; Lockwood and Shah, 1981), have also been used. The heat absorbing walls were modeled as constant-temperature walls (Yin et al., 2002; Zhou et al., 2002). The heat exchangers were modeled either by the porous volume approach (He et al., 2007; Yin et al., 2002) or by the constant-temperature double-sided walls approach (Yin et al., 2002).

The energy balance for the discrete phase can be written as

$$M_p Cp_P \frac{dT_P}{dt} = \left(f_{heat} \frac{dM_c}{dt} H_{rxn} \right) + Q_{rad} + Q_{conv} - \sum \frac{dM_v}{dt} H_{fg} \qquad (4.47)$$

Here, Cp_p, H_{rxn}, Q_{rad}, and Q_{conv} are the particle specific heat, heat of char oxidation reaction, and radiative and convective heat transfer, respectively. The f_{heat} is the fraction of heat absorbed by the coal particle during char oxidation, H_{fg} is the latent heat of evaporation of the volatile/moisture, and $\frac{dM_v}{dt}$ is the rate of evaporation of volatile/moisture.

The particle radiative heat transfer can be written as

$$Q_{rad} = \varepsilon_P \sigma A_p (T_R^4 - T_P^4) \qquad (4.48)$$

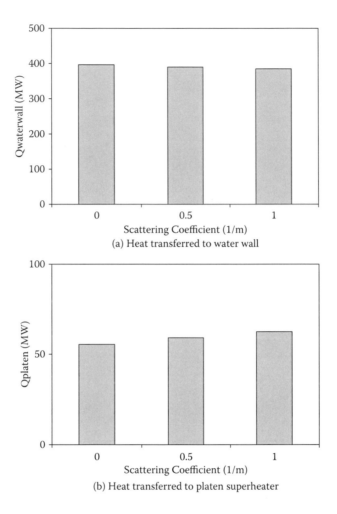

FIGURE 4.5
Influence of scattering coefficient on heat transfer in a typical 210-MWe PC fired boiler: (a) heat transferred to water wall and (b) heat transferred to platen superheater.

where ε_p is the particle emissivity, σ is the Stefan–Boltzmann constant $(= 5.67 \times 10^{-8}\,W/m^2K^4)$, T_R is the radiation temperature $\left[\left(\frac{I}{4\sigma}\right)^{1/4}\right]$, and h is the heat transfer coefficient.

Backreddy et al. (2006) have proposed a value of 0.9 for particle emissivity. Asotani et al. (2008) and Lockwood et al. (1986) have proposed the following correlation: $\varepsilon_p = 0.4\,UBC + 0.6$ for estimating particle emissivity (where UBC is unburned carbon).

The convective heat transfer can be written as

$$Q_{conv} = hA_P(T_g - T_P) \tag{4.49}$$

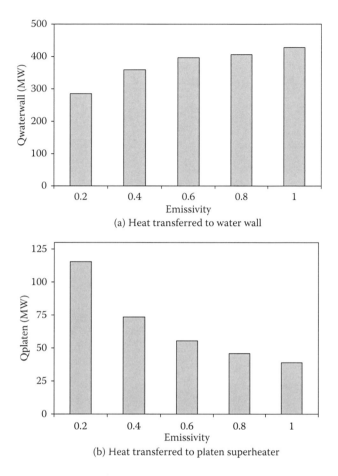

FIGURE 4.6
Influence of wall emissivity on heat transfer in a typical 210-MWe PC fired boiler: (a) heat transferred to water wall and (b) heat transferred to platen superheater.

The heat transfer coefficient h was evaluated using the correlation of Ranz and Marshall (1952) as

$$\frac{h d_p}{k_g} = 2 + 0.6 (\mathrm{Re}_p)^{1/2} (\mathrm{Pr})^{1/3} \qquad (4.50)$$

Further details of heat and mass transfer correlations can be found in Appendix 4.3 of Ranade (2002).

This completes the formulation of the basic governing equations for simulating processes occurring in a typical PC fired boiler. Before we discuss simulations using these basic governing equations, it will be useful to discuss two additional topics in the formulation of model equations that are quite relevant in practice, namely (1) simulating the formation of pollutants (NO_x and SO_x) and (2) deposition of particles on internal walls.

4.1.4 Formation of NO_x and SO_x

The combustion of coal in PC fired boilers is accompanied by the formation and emission of gases harmful to the environment, such as NO_x and SO_x. The volume of sulfur in exhaust gases is strictly correlated with the sulfur content of the combusted coal, unlike NO_x, which may form from the inherent nitrogen content of the air fed to the PC fired boiler. A detailed discussion of the modeling of NO_x and SO_x formation in PC fired boilers is outside the scope of this book. In this section we briefly discuss key issues and cite some references that may be consulted for obtaining further details.

Most of the sulfur contained in fuels (more than 95%) is oxidized to form SO_2. This may further react with hydroxyl radicals and form SO_3. The SO_3 may then react with water drops in clouds and result in acid rain. The mechanisms of the release of sulfur compounds from coal and subsequent oxidation to SO_2 are influenced by many factors, including coal type, content and characteristics of the sulfur-containing compounds, and combustion conditions. In general, organic sulfur compounds are less stable than inorganic sulfur compounds. Organic sulfur usually is released during the devolatilization process, forming sulfuric species. However, a certain fraction of the organic sulfur remains trapped in the char and is therefore released subsequently during char burnout (see Muller et al. (2013) and references cited therein for more details). Significant efforts have been made to develop computational models to simulate SO_x formation and thereby evolve ways for possible reductions in the formation of SO_x. It should be noted that the sulfur content in Indian coal is much lower (<1%) compared to that in coal in the United States (Visuvasam et al., 2005). Because the focus of this book is on modeling PC fired boilers using high-ash-containing Indian coal, further discussion on modeling SO_x formation is not included here. More details may be found in Krawczyk et al. (2013) and references cited therein.

Unlike SO_x formation, which is related to the sulfur content of coal, NO_x can form via two possible sources: (1) reaction of nitrogen in air with oxygen to form NO_x (which is temperature dependent) and (2) oxidation of fuel nitrogen present in volatile and char (fuel dependent). There is increasing emphasis on reducing pollutants such as NO_x. Several ways, such as staging of the combustion air or low NO_x burners, are currently practiced by boiler manufacturers to reduce NO_x emissions. Computational models play a very important role in simulating and thereby evolving ways for reducing NO_x emissions. Several mechanisms and models have been proposed to represent the formation of thermal NO_x and fuel NO_x (see Dean and Bozzelli (2000) and references cited therein). Some of these aspects are discussed briefly here.

Formation of NO_x via a thermal mechanism (Zeldovich mechanism) is important when the local temperature in the flame increases to more than 1,800K. The rate expressions and kinetic parameters for thermal NO_x formation have been extensively studied and reported in the published literature (see, for example, Hanson and Salimian (1984)). The available information indicates that formation of a nitrogen atom is the rate-limiting step. Therefore, it is generally assumed that the rate of production of NO is the same as the rate of generation of nitrogen atoms. Usually, the quasi-steady-state assumption is invoked to simplify rate expressions. These quasi-steady-state-based rate expressions involve concentrations of radicals that may be correlated with molecular concentration by assuming that

- Equilibrium for [O] and neglecting [OH] radicals
- Partial equilibrium for [O] and [OH]

The equilibrium approach may cause as much as a 25% under-prediction of NO_x values in the flame zone. The contribution of thermal NO_x is generally less than 10% to 15% of the total NO_x formed (Stanmore and Visona, 2000). The remainder typically originates from fuel.

Most of the nitrogen found in coal is embedded as aromatic nitrogen compounds such as pyridine (C_5H_5NH), pyridone ($C_5H_4NH(O)$), and pyrrole (C_4H_4NH). The low-ranked (less mature and high oxygen containing) coals generally contain pyridone as a major compound; and as the coal becomes more mature, the pyridone loses its oxygen and transforms into pyridine (Speight, 2013). During this thermal transformation, the nitrogen present in the parent coal is distributed into volatile matter and char. It is reported that the volatile nitrogen accounts for nearly 45% to 80% of the total NO_x emissions for fuel-lean conditions and ~10% for fuel-rich conditions (Diez et al., 2008; Spitz et al., 2007). Distribution of the nitrogen and the types of nitrogen compounds affect NO_x formation in PC fired boilers (Speight, 2013; Davidson, 1994). Other factors such as thermal history and an oxygen-rich environment due to improper furnace and burner design may promote NO_x formation in the boiler.

The volatiles may undergo secondary pyrolysis to release nitrogen (N_2), hydrogen cyanide (HCN), ammonia (NH_3), and amine compounds that may oxidize to form nitric oxide (NO) and nitrous oxide (N_2O). Fuel nitrogen transformed into HCN and NH_3 will contribute to major NO_x formation. Low-rank coal produces ten times more NH_3 than HCN, while bituminous coal produces only HCN (Winter et al., 1996; Naruse et al., 1995). The formation of HCN or NH_3 in the volatile matter is a function of coal rank and also the oxygen available in the vicinity. An oxygen-lean environment promotes N_2 formation while an oxygen-rich environment promotes NO_x formation. Spinti and Pershing (2003) have suggested that the char-N conversion to NO_x

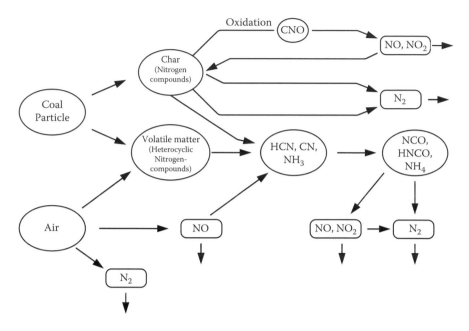

FIGURE 4.7
Schematic representation of NO_x formation during coal combustion.

for a typical PC fired boiler is 50% to 60% for low-rank coals and 40% to 50% for bituminous coals. Williams et al. (2001) have proposed that char-N amounts to about 80% of the total NO formed. In contrast to this, Rostam-Abadi et al. (1996) have indicated a 20% to 30% contribution from char-N. A typical representation of the NO_x process from pulverized coal is shown in Figure 4.7.

Fuel NO_x forms via the oxidation of volatile-nitrogen and char-nitrogen and may proceed according to one of the following routes:

- Char-N and volatile-N form HCN or NH_3, or both.
- Char-N directly oxidizes to NO due to pore reactions, and volatile-N forms HCN or NH_3, or both.

Typically, conversion of char-N to NO_x is slower compared to the conversion of volatile-N. Assumptions about the relative formation of HCN and NH_3 are important because this affects NO_x formation. For low-rank coal, NH_3 is the primary product, while for high-rank coal (e.g., bituminous coal), HCN is the major product.

The brief discussion in this section can be used to select appropriate models for representing NO_x and, if necessary, SO_x formation in PC fired boilers. The formulated model equations can be assimilated with the species balance equations discussed previously by appropriately selecting effective rate expressions. The references cited in this section should be consulted for further details.

4.1.5 Particle Deposition on Walls and Heat Transfer through the Deposit Layer

Deposition of the ash on the surface of the heat exchanger is one of the critical problems for any solid fired combustion system such as a PC fired boiler because it impacts the heat transfer efficiency of the system. Generally, soot blowers are used to remove these depositions from the surface of the heat exchanger. Steam is used to blow the depositions. High steam consumption, fire-side corrosion of metal tubes, frequent maintenance, and sometimes unwarranted shutdowns are key operational issues in PC fired boilers. Such problems become more severe for high-ash-containing coal. The problem is even more severe with typical Indian coals because the ash content in these coals is more abrasive due to the high quartz (silica) content. In typical cases where coal is co-fired with biomass, fouling issues are quite critical. Biomass ash contains low-melting salts that may lead to severe fouling issues in the boiler. Salts such as those based on Na, K, and S (e.g., Na_2SO_4, K_2SO_4) vaporize above 750°C and condense on steam-cooled tube surfaces where they form sticky deposits. Particle deposition on any heat transfer surface will significantly influence heat transfer rates. It is therefore important to develop appropriate models to capture possible particle deposition in PC fired boilers and to account for the influence of particle deposition on effective heat transfer in such boilers.

Two terms quite commonly used are *slagging* and *fouling*. Slagging mostly occurs in the radiant furnace zone of PC fired boilers where the coal particle temperature exceeds the melting point of the ash. The deposition surface temperature influences the microstructure of the deposit (see Zhou et al., 2014). The melted ash forms a sticky viscous mass and deposits over the water walls of the furnace. Therefore, the slagging depositions are mostly viscous sticky liquid layers on the tube surfaces of the furnace. The fouling occurs in the upper section of the furnace where superheaters and reheaters are suspended in the flow path of the flue gas. The flue gas entering this zone is at a comparatively lower temperature (<1000°C). The condensable species in the flue gas condense out on the heat exchanger surfaces (<600°C) and form dry deposits, unlike slag. Typical photographs illustrating fouling in PC boilers are shown in Figure 4.8.

There are differences in the deposition within the furnace and on the superheater surfaces in terms of their composition and thickness of deposit layer. The thicknesses of deposits on the superheater are reported in the range of 25 to 30 mm over a period of 6 to 9 months. The first 25 mm grows quickly over the superheater surface. The rate of deposition then slows down and only a few millimeters of growth is observed in the next 6 months (Tomeczek

FIGURE 4.8 (See Color Insert.)
Fouling observed in PC fired boilers: deposit build-up on superheater after 1 week of co-firing (coal + straw). Reprinted from *Progress in Energy and Combustion Science, 31* (5–6) Zbogar, A., Flemming, J.F., Jensen, P.A. and Glaborg, P., Heat Transfer in Ash Deposits: A Modelling Tool Box 261–269 (2005) with permission from Elsevier.

et al., 2004). The composition of the deposits on superheater surfaces mainly contains $Al_2O_3 \cdot 2SiO_2$, SiO_2, Fe_2O_3, CaO, and MgO. Alkalis such as Na_2SO_4, K_2SO_4, KCl, and NaCl account for less than 2%. There is almost no carbon in these deposits. The growth of deposits at the end of 3 months of operation in a typical PC fired boiler is shown in Figure 4.9.

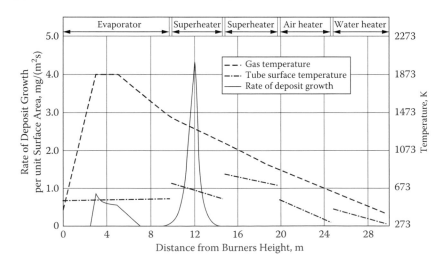

FIGURE 4.9
Deposit growth on PC boiler tubes at the end of 3-month operation. Reprinted from *Fuel*, 83(2), Tomeczek, J. Polugniok, H. and Ochman, J., Modelling of deposit formation on heating tubes in pulverized coal boilers, 213–221, (2004), with permission from Elsevier.

There are four main modes in which particle deposition occurs on the surface of the wall:

1. Inertial impaction due to inertia and drag (particle size >10 μm)
2. Thermophoretic attraction (particle size < 10 μm)
3. Condensation (particle size < 1 μm)
4. Heterogeneous reaction between particle and wall surface (particle size <1 μm)

These parameters are affected by the hydrodynamics of the furnace and turbulent interaction (eddy and particle) between gas and solid particles. The overall approach of modeling particle deposits is shown schematically in Figure 4.10.

The rate of particle deposition (N_p) is given as

$$N_p = J \cdot \eta_i \cdot \eta_s \tag{4.51}$$

where J is the flux of particles, η_i is the impact efficiency, and η_s is the sticking efficiency.

The probability of particle impact with the heat exchanger surface is called the *impact efficiency*. Larger particles have higher particle relaxation times and hence will have a larger probability of colliding with the surface. The small particles with smaller particle relaxation times will follow gas flow and have a lower probability of impact with surfaces. The impact efficiency is usually expressed as a function of the Stokes number, which is the ratio of the particle relaxation time (τ) to particle flow characteristic time. Detailed

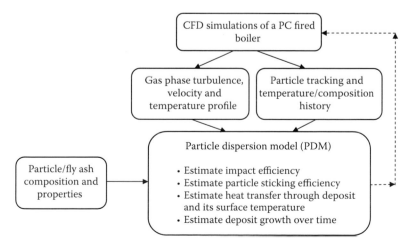

FIGURE 4.10
Overview of modeling of the ash deposits on the surface of tube heat exchanger.

expressions for estimating the impact efficiency can be found in Huang et al. (1996) and Baxter (1990).

All the particles that collided with the walls of internal heat exchangers may not deposit on the walls. It is therefore essential to consider the probability of a particle sticking to the surface. This depends on particle properties such as the temperature of the particle, composition, angle of approach, and kinetic energy, as well as tube properties such as temperature, roughness, and composition of any already-coated ash layer. To quantify this probability of sticking, the critical viscosity (of the slag) based model is generally adopted. The probability of a particle sticking to the surface is equal to the ratio of critical viscosity to particle viscosity with a maximum value of 1. The typical value of the critical viscosity of slag is 10^4 Pa-s (Hao et al., 2002; Huang, 1996; Richards et al., 1992). A few researchers have adopted a value for the critical viscosity of 10^5 Pa-s (Lee and Lockwood, 1999; Wang and Herb, 1997), and others have suggested a broad range of values from 10^4 to 10^8 Pa-s (Walsh et al. (1990) and Srinivasachar et al. (1990)). Various expressions used to estimate particle viscosity are summarized in Table 4.3.

Another approach for estimating the sticking efficiency was proposed by Tomeczek (2004); it is based on the particle temperature and the particle melting temperature. Based on estimations of impact efficiency and sticking efficiency along with the flux of the particle simulated using the CFD model, the effective rate of deposition can be estimated as a post-processing of the CFD results.

Once the thickness of the deposit is estimated, the influence of the deposit on heat transfer must be accounted for in the CFD model. Heat transferred through ash deposition is the net result of the heat convected and radiated

TABLE 4.3

Summary of Estimation of Particle Viscosity

Ref.	Correlations for η (Poise)	Remarks
Watt and Fereday (1969); Huang (1996);	$\log(10\mu) = \dfrac{10^7 m}{(T*150)^2} + c$	$m = 0.00835[SiO_2] + 0.00601$ $[Al_2O_3]-0.109$ $c = 0.0415[SiO_2] + 0.0192\,[Al_2O_3] + 0.0276$ [eq. Fe_2O_3] + 0.016 [CaO] -3.92 $[SiO_2] + [Al_2O_3] + [eq.\ Fe_2O_3] + [CaO]$ $= 100$ wt%
Urbain (1981,1982)	$\ln\eta = \ln A + \ln T + 10^3 B/T - \Delta$	Equation for slag viscosity. Depending upon the silica in the slag the values of constant varies.
Lucas (2001)	$\ln\eta = 4.468\left(\dfrac{s}{100}\right)^2$ $+1.265\left(\dfrac{10^4}{T}\right)-7.44$	S: silica ratio; T is in K

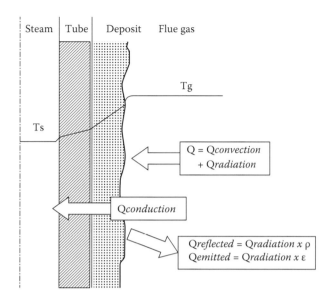

FIGURE 4.11

Heat transfer to and through ash deposits. Reprinted from *Progress in Energy and Combustion Science*, 31 (5-6), Zbogor, A., Flemming, J.F., Jensen, P.A. and Glaborg, P., Heat Transfer in Ash Deposits: A Modelling ToolBox, 261–269 (2005), with permission from Elsevier.

to the deposit and the heat radiated and emitted by the deposit (shown schematically in Figure 4.11):

$$Q_{cond} = Q_{conv} + Q_{rad} - Q_{ref} - Q_{emm} \tag{4.52}$$

where Q_{cond} is the heat conducted through the porous deposit to the heat exchanger surface, Q_{conv} is the heat convection to the deposit from the flue gas ($h(T_g - T_d)$), Q_{rad} is the heat radiated from the gas and other emitting bodies ($\epsilon \cdot \sigma \cdot T_g^4$), Q_{ref} is the radiation reflected back from the deposit ($\varphi \cdot \epsilon \cdot \sigma \cdot T_g^4$), and Q_{emi} is the heat emitted by the deposit ($\epsilon_d \cdot \sigma \cdot T_d^4$). Wherein, h is the convective heat transfer coefficient between the flue gas and the deposit, T_g is the flue gas temperature, T_d is the temperature of the deposit, ϵ is the emissivity of the flue gas or other emitting bodies, σ is the Stefan–Boltzmann constant, φ is the reflectivity of the deposit, and ϵ_d is the emissivity of the deposit.

If the deposits on the tube surface are of cylindrical shape (and uniform along the periphery), the heat of conduction for a cylindrical pipe can be written as

$$Q_{cond} = \frac{T_d - T_t}{\frac{d_d}{2 \pi k_d} \ln \frac{d_d}{d_t}} \tag{4.53}$$

where T_t is the tube metal temperature, d_d is the outer diameter of the deposit, d_t is the outer diameter of the tube, and k_d is the conductivity of the deposit.

The temperature of the deposition (T_d) can be estimated using Equations (4.52) and (4.53).

The conductivity and emissivity of the deposit depend on the particle size, composition, and thickness of the deposition. An extensive review of particle deposition and heat transfer has been presented by Zbogar (2005) and may be referred to for more details.

This completes the formulation of model equations. The next step is to use appropriate numerical methods to solve the model equations and then apply the computational model to simulate PC fired boilers. This is illustrated here with an example of a typical 210-MWe PC fired boiler.

4.2 CFD Simulations of a PC Fired Boiler

In recent years, significant efforts have been made to carry out CFD simulations of different types of PC fired boilers (wall or corner fired). CFD simulations have been performed to understand key aspects of the flow processes occurring in PC boilers and also gain insight, which quite often measurements alone may not practically provide. Most of the previous CFD studies were simulated gas flow field, temperature and species distribution, and particle trajectories within the boiler; see, for example, Yin et al. (2002) and Fan et al. (2001, 1999). Efforts have also been made to simulate NO_x production in PC fired boilers (and to evaluate alternatives for reducing the same; see, for example, Backreedy et al. (2005), Stanmore et al. (2000), and Xu et al. (2000)).

There were attempts to predict the temperature deviation that occurs in the upper furnace zone (crossover pass) near the superheaters (Yin et al., 2002; Xu et al., 1998). It has been shown that residual swirl is the primary cause of the observed temperature deviation in the boiler. Ash deposition on the heat exchangers is a common phenomenon in PC fired boilers, and models have been developed for simulating the ash deposition on the heat exchanger walls (Zhou et al., 2002; Fan et al., 2001). The fuel-to-air ratio plays an important role in determining the overall oxygen concentration and temperature profiles, residence time of the particles in the furnace, unburned char remaining in ash, and NO_x formation. Operational parameters such as a reduction in boiler heat load from standard operating conditions have an influence on boiler performance. Xu et al. (2001) have performed simulations to understand the effect of heat load on boiler performance for a 300-MWe wall fired boiler. More recently, Belosevic et al. (2008) performed numerical studies to understand the effect of operating conditions on the performance of a 350-MWe wall fired boiler (for lignite coal). Here we illustrate application of the CFD model discussed in the previous section to a 210-MWe tangentially fired PC boiler. The methodology, however, can be extended in a straightforward manner to any PC boiler, including current-generation PC fired boilers.

Key steps in the application of the CFD model for simulating PC fired boilers and a systematic methodology for it are discussed in Chapter 6. It is recommended to read that chapter before going through the remainder of this chapter. It may be necessary to read these two chapters iteratively to facilitate a better understanding of the salient aspects of applying computational models to PC fired boilers. In this chapter, key steps discussed in Chapter 6 are illustrated with the case of a typical 210-MWe PC fired boiler. The geometry considered here broadly follows the configurations used by many of the Indian utility companies. The considered PC boiler is fired with medium-volatile, high-ash-content subbituminous type coal. The commercial CFD solver FLUENT® version 6.2 (ANSYS, Inc., Canonsburg, Pennsylvania) was used to solve the mass, energy, and momentum governing equations for gas and particles. Before the results of an illustrative application of the CFD model to a considered PC boiler, some comments on grid generation and simplifying considerably the geometry of a PC boiler are included in the next section. The illustrative results obtained by applying the computational model are then discussed for the following cases:

- Cold air velocity tests
- Base case simulation
- Influence of key design and operating parameters
  - Burner tilt
  - Excess air
  - Turndown (coal feed rate)
- Coal blends

4.2.1 Numerical Simulation and Simplifying the Geometry of PC Fired Boilers

The model equations discussed in Section 4.1 are strongly nonlinear and coupled partial differential equations. These model equations must be solved numerically. In general, numerical solution of the governing transport equation replaces continuous partial differential equations by discrete information contained in a set of algebraic equations. The set of algebraic equations (called discretized equations) involving the unknown values of dependent variables is derived from the governing partial differential equations. Some assumptions about how the unknown dependent variables change between grid points are necessary for such derivation. Generally, piecewise profiles are assumed that describe variation over a small region around the grid point in terms of values at that grid point and the surrounding grid points. To facilitate this, the solution domain is divided into a number of computational cells (the process is called *grid generation*) so that a separate profile assumption can be associated with each computational cell. Various different methods are available for deriving the discretized equations. Most commercial CFD tools applicable to turbulent, multiphase, and reacting flows are based on the finite volume method (Patankar, 1980; Ranade, 2002). The discretized equations formulated using the finite volume method are then iteratively solved using suitable algorithms for treating pressure–velocity coupling. Special algorithms are needed to solve multiphase flows. The radiation transport equation also demands different and special treatment because of its different intrinsic character. It will not be possible to discuss the details of numerical solution here. The reader is referred to the book by Ranade (2002) and references cited therein. The approach of numerical solution and some of the key steps are highlighted here for the sake of completeness.

After selecting the numerical method, it is necessary to generate an appropriate grid, that is, a discrete representation of the solution domain and discrete locations at which variables are to be calculated. Two types of grids, namely structured and unstructured grids, are used in practice. In a structured grid, there are families of grid lines following the constraint that grid lines of the same family do not cross each other and cross each member of the other families only once. The position of a grid point within the solution domain is, therefore, uniquely identified by a set of three indices. It is thus logically equivalent to a Cartesian grid. The properties of a structured grid can be exploited to develop very efficient solution techniques. One major disadvantage is the difficulty in controlling the grid distribution. In a structured grid, concentration of grid points in one region for more accuracy may unnecessarily lead to small spacing in other parts of the solution domain. A block-structured grid is used to eliminate or reduce this disadvantage. In a block-structured grid, the solution domain is divided into a number of blocks that may or may not overlap. Within each block, a structured grid is defined. This kind of grid is more flexible as it allows local (block-wise) grid refinement. For very complex geometries, unstructured grids, which can fit an arbitrary solution domain boundary, are used. In this case, there is no restriction on the shape of the control volume and number of neighboring nodes. Triangles and quadrilaterals in two dimensions (2D) and tetrahedral or hexahedral in three dimensions (3D) are the most widely used grid shapes in practice. Unstructured grids can be refined locally and allow more control of the variation of aspect ratio, etc. The advantage of flexibility is often offset by the disadvantage of the irregularity of the data structure. The solvers for algebraic equation systems of unstructured grids are generally slower than those for structured grids. Several excellent texts on grid generation are available; see, for example, Thompson et al. (1985) and Arcilla et al. (1991). Some commercial grid generation tools are summarized by Ranade (2002), along with a few tips on generating suitable grids for simulating complex industrial flow processes.

4.2.1.1 Grid Generation for a PC Fired Boiler

A PC boiler is a huge structure mainly comprised of tubes. The overall structure and several internal heat exchangers were discussed in Chapter 1. Appropriate modeling of internal heat exchangers such as the superheater, reheater, and economizer is important in predicting the flue gas temperature profile, velocity profile in the boiler, and the heat transferred to each unit. Each of these internal heat exchangers has many tubes, with each tube extending to several tens of meters. The characteristic size of an individual tube is very small compared to the overall size of a PC fired boiler. If an individual tube must be resolved while generating the grid, the total number of computational cells will escalate very rapidly, and the simulation may become intractable. Therefore, the internal heat exchangers are often represented by a porous block that has characteristics similar to those of an internal heat exchanger. This approach simplifies the overall geometry and therefore the process as well as the quality of grid generation (see Figure 4.12).

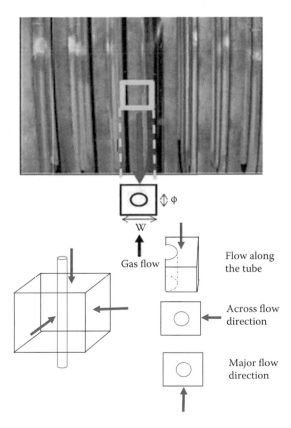

FIGURE 4.12 (See Color Insert.)
Representation of actual tube bundle of internal heat exchanger as a porous volume.

The porosity, pressure drop, and heat transfer characteristics of the porous block must be defined in such a way as to mimic the real characteristics of the internal heat exchanger. The characteristics of an internal heat exchanger can be obtained either from experiments or from separate CFD models. Here we illustrate the use of separate CFD models to characterize the internal heat exchanger. The characteristics of the internal heat exchangers are then specified to the porous blocks representing the heat exchangers in the overall CFD model for a PC boiler.

4.2.1.2 Characteristics of Porous Block Representing Internal Heat Exchangers

Flue gas, after leaving the furnace section of the boiler, approaches the tube bundles suspended in the flow path. The flow area within each tube bundle is partially occupied by the tubes. Hence, the velocity of the flue gas increases when it passes through the channel between two tubes. To capture this change in velocity, information about the volume occupied by the heat exchanger tubes is required, and this information can be estimated based on the outer diameter (OD) of the tube, its length, and the number of tubes present in each tube bundle. Based on such information, the volume porosity (ratio of the volume occupied by the fluid to the total volume of the tube bundle) can be estimated and specified for all the porous blocks representing the internal heat exchanger. This methodology takes care of the flue gas velocity within the tube bundle. It is possible that there may be variation in this porosity value within the tube bundle, especially at the lateral ends near the walls where the tube-to-wall distance may be different. For such cases, multiple porous blocks can be used to represent internal heat exchangers, and appropriate values of porosity may be set for these porous blocks.

Based on their geometric configuration, such as tube OD, pitch between the tubes, etc., internal heat exchangers offer an extra pressure drop to flow. The porous block representing these internal heat exchangers should mimic these characteristics. This can be achieved by specifying an appropriate momentum sink within the porous block. Typically, such a momentum sink is defined by specifying suitable viscous and inertial resistances as

$$\frac{dP}{dz} = C_0 \mu U + C_2 \frac{1}{2} \rho U^2 \tag{4.54}$$

where $\Delta P/L$ (or dP/dz) is the pressure drop per unit length (Pa/m)/volumetric momentum sink, C_0 is the coefficient of viscous resistance (1/m²), C_2 is the coefficient of inertial resistance (1/m), U is the gas velocity (m/s), and μ is the gas viscosity (kg/ms).

The C_0 and C_2 are functions of geometric parameters such as tube OD, the pitch between two tubes, pitch pattern, etc. A unit cell approach can be used to obtain suitable values of C_0 and C_2 (see Gunjal et al. (2005a) and Ranade et al. (2011) for more details). In the unit cell approach, end effects are ignored, and a periodic section of a tube bundle is considered for simulating flow and heat transfer. Flow through closely packed tubes has been observed to become periodic. This means that flow around any tube in the bundle may be assumed to be same, and therefore information obtained from the periodic unit cell can be used for the entire tube bundle after appropriate scaling. The concept of the unit cell is illustrated in Figure 4.12(b), which shows a 3D unit cell with a single tube. The three arrows indicate the direction in which the flow will approach the unit cell/tube bundle.

To illustrate this approach, some representative results are presented here. Periodic unit cells were formulated to represent configurations of all the internal heat exchangers in a typical 210-MWe PC fired boiler. The flow and heat transfer in such a periodic unit cell are then simulated over a range of velocities in all three directions. Typical values of the simulated pressure drop for various values of superficial velocity are shown in Figure 4.13.

The values of the viscous and inertial resistance coefficients (C_0 and C_2) can be obtained by fitting Equation (4.54) to the simulated values of pressure drop. The values obtained for porous blocks representing typical internal heat exchangers of PC fired boilers with such a unit cell approach are listed in Table 4.4.

The unit cell approach also allows one to simulate periodic heat transfer. Heat transfer simulations of such unit cells are usually carried out with a constant wall temperature boundary condition. The wall temperature can be specified as the average temperature of the inlet and outlet of steam/water for each tube bundle. Typical simulated results of simulations of heat transfer in the form of temperature contour plots are shown in Figure 4.14. Heat

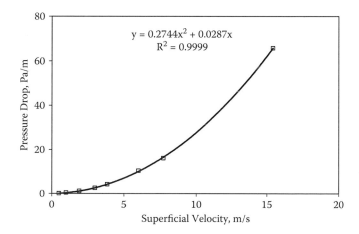

FIGURE 4.13
Simulated results for pressure drop across a typical platen superheater.

TABLE 4.4

Typical Pressure Drop and Nusselt Number Coefficients for Heat Exchangers

| | | Pressure Drop Coefficients $\dfrac{dP}{dz} = C_0 \mu v + C_2 \dfrac{1}{2}\rho v^2$ | | Nusselt Number Coefficients $Nu = c\, Re^m\, Pr^{0.33}$ | |
| | | Approximate Viscous Resistance Coefficients | Approximate Inertial Resistance Coefficients | | |
Sr. No.	Heat Exchanger	$(C_0,\ 1/m^2)$	$(C_2,\ 1/m)$	c	m
1	Platen SH	$1.5\text{--}2 \times 10^3$	0–1	0.4	0.61
2	Front and Rear Reheater	-	1–2	0.24	0.63
3	Final SH	$1.5\text{--}2 \times 10^3$	8–10	0.42	0.61
4	LTSH	$1.5\text{--}2 \times 10^3$	8–10	0.42	0.61
5	Lower and Upper Economizer	$5\text{--}5.5 \times 10^4$	8–10	1.27	0.52

transfer coefficients can be calculated from the simulated results. The values of the heat transfer coefficient obtained from such unit cell simulations are usually correlated in the form of the Nusselt number as

$$Nu = c\, \mathrm{Re}^m\, \mathrm{Pr}^{1/3} \tag{4.55}$$

$$Nu = \frac{hD_t}{k} \tag{4.56}$$

where h is the local heat transfer coefficient, D is the characteristic dimension (outer diameter of the tube), and k is the conductivity of the flue gas. An illustration of fitting the simulated values of Nusselt numbers using Equation (4.55) is shown in Figure 4.15. The fitted values of c and m for a wide range of Reynolds numbers (Re) for typical internal heat exchangers of PC fired boilers are listed in Table 4.4.

Once the resistance and heat transfer coefficients of the tube bundle are obtained from such a unit cell approach, user-defined functions (UDFs) for

1550 K

673 K

FIGURE 4.14 (See Color Insert.)
Simulated temperature distribution across a tube of a typical platen superheater.

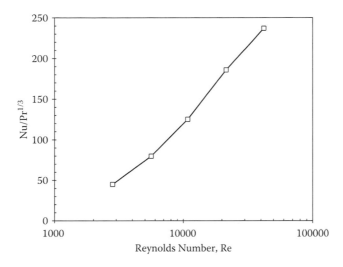

FIGURE 4.15
Simulated heat transfer coefficient for a typical platen superheater.

providing the appropriate momentum and heat transfer sink can be formulated in a straightforward manner. Almost all commercial CFD solvers allow users to include such user-defined sources/sinks while formulating a computational model.

With the use of the porous block approach for representing internal heat exchangers, geometry modeling and grid generation for a PC fired boiler become drastically simplified. A typical 210-MWe PC fired boiler with overall dimensions of 25 m × 14 m × 52 m (see Figure 4.16) was considered for this illustration. The furnace (consists of furnace and nose), crossover pass (consists of platen superheater, reheater, and final superheater), and rear pass (consists of LTSH, lower and upper economizer) were considered in the solution domain. Four operating fuel air (FA) burners (injecting primary air and coal) and five auxiliary air (AA) burners (injecting secondary air) were considered at each corner of the furnace (arranged alternatively).

The air and coal are projected from the nozzles at an angle of α or β degrees with respect to the Y axis at a particular Z plane (Figure 4.16(b)). The burners were represented as flat surfaces.

The grid has a significant impact on the rate of convergence (or even lack of convergence), solution accuracy, and CPU (central processing unit) time required. The mesh quality for good solutions depends on grid density, adjacent cell length/volume ratios, and skewness. The mesh density should be high enough to capture all relevant flow features. The geometry was meshed with hexahedral cells and tetrahedral mesh at some parts. Figure 4.17 shows the generated grid at the cross-section of the furnace. Initially, the geometry was meshed with 0.87 million cells. The grid was further refined to generate

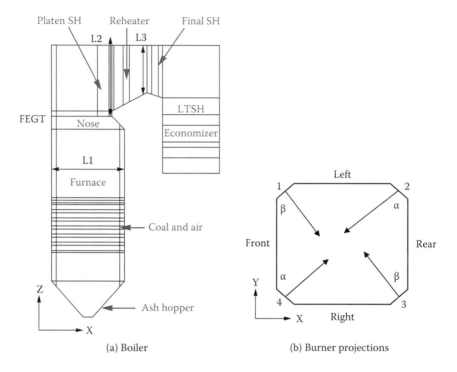

(a) Boiler

(b) Burner projections

FIGURE 4.16
Schematic of 210-MWe tangentially fired PC boiler: (a) boiler and (b) burner projections.

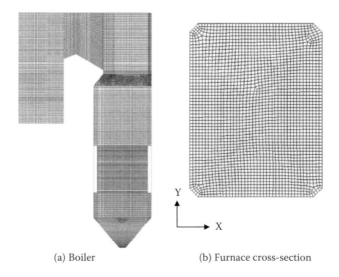

(a) Boiler

(b) Furnace cross-section

FIGURE 4.17
Computational grid used to illustrate the application of a CFD model for a PC boiler: (a) boiler and (b) furnace cross-section.

1.4 million and 2.0 million cells to evaluate the influence of grid size on the predicted results.

This computational grid and resistance coefficients obtained previously were used to formulate a computational model. As a first step, the application of such a computational model to simulate cold air velocity tests is discussed in the following section.

4.2.2 Simulation of Cold Air Velocity Tests (CAVTs)

Single-phase air flow simulation was performed with different computational cells to quantify the influence of computational cells on the predicted results. The gas flow inlets for fuel air (FA) and auxillary air (AA) were defined as the velocity inlet. The outlet was specified as the pressure outlet. A no-slip condition was specified at all the walls. The predicted velocity magnitude profiles are shown in Figure 4.18 for three different cases (with 0.87, 1.4, and 2.0 million cells). It can be seen that the predicted results indicate a small influence due to the number of computational cells (Figure 4.18). It can be seen that no significant change in velocity magnitude can be observed between the results predicted with 1.4 and 2.0 million cells. Therefore, all subsequent simulations were carried out with 1.4 million computational cells (i.e., 1,465,013 grid cells).

Because the flow in a coal fired boiler is turbulent, it is important to select an appropriate turbulence model. Two-equation turbulence models were found to be most appropriate and widely used in the literature for simulating PC fired boilers. Numerical experiments were performed to understand the influence of different two-equation turbulence models, namely the (1) standard k–ε model, (2) renormalization group (RNG) version of the standard k–ε model, and (3) realizable k–ε model. The simulated results are shown in Figure 4.19. The predicted results at L3 in the crossover pass (Figure 4.19(a)) indicate no significant influence of turbulence models such as standard, renormalization group, or realizable on the predicted velocity profiles. The predicted results of different turbulence models were also compared in the furnace zone where flow is swirling in nature. As a sample of such a comparison, predicted velocity profiles with three turbulence models at L1 are shown in Figure 4.19(b).

It can be seen that predicted results of the standard k–ε model and the RNG k–ε model of turbulence agree with each other quite well. The results predicted with the realizable k–ε model show some deviation from these results. There is no adequate data available to discriminate between these models. The RNG k–ε model was used for subsequent simulations. The model was then extended to simulate a typical PC boiler.

4.2.3 Simulation of a Typical 210-MWe High-Ash Coal Fired Boiler

The same geometry and computational grid used for simulating CAVTs is used for simulating a typical 210-MWe PC fired boiler. Flow-related boundary

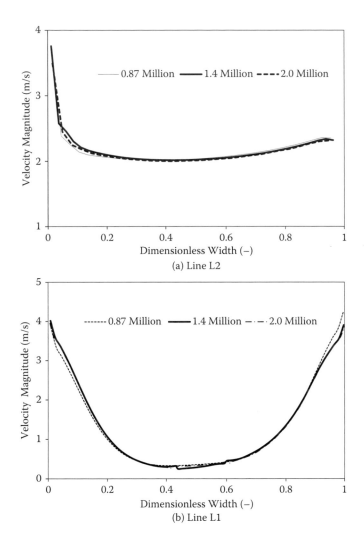

FIGURE 4.18
Influence of the number of computational cells on simulated profiles of velocity magnitude:
(a) Line L2 and (b) Line L1.

conditions were the same as those used for the CAVT. Formulation of various additional boundary conditions and other aspects of defining coal particles and reaction parameters are briefly discussed below.

For the discrete phase, the coal particles were injected as a surface injection. The reflect condition was specified for the particles at the wall, and the escape condition was specified at the outlets. Particles were modeled as discrete phase with particle size distribution (PSD) specified in terms of Rosin–Rammler (RR) parameters (see Equation (3.12) in Chapter 3). Specific operating

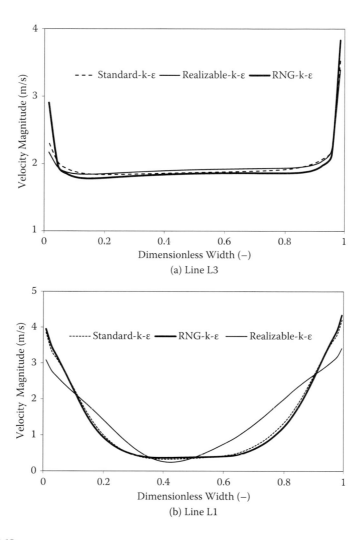

FIGURE 4.19
Influence of turbulence models on simulated profiles of velocity magnitude: (a) Line L3 and (b) Line L1.

conditions used for carrying out the simulations of a base case are listed in Table 4.5.

Appropriate momentum and heat sink terms were added to represent all the internal heat exchangers. The emissivity of heat exchanger tube bundles was specified as 0.6, and the tube wall temperature was estimated from the average value of the input and output temperatures of the steam. The emissivity of the water walls was specified in the range of 0.6 to 0.8. Appropriate kinetic parameters for devolatilization and char oxidation were obtained

TABLE 4.5

Operating Conditions Used for Simulating a Typical PC Fired Boiler

Operating Heat Load (MWe)		210			168 (20% Turndown)	126 (40% Turndown)
Excess Air (%)	20	10	5		20	20
Fuel air mass flow rate (kg/s); (Temperature 350K)	75	69	66		60	45
Auxiliary air mass flow rate (kg/s); (Temperature 553K)	149	135	129		119.2	89.4
Coal flow rate (kg/s)[a]; (Temperature 350K)			39		31.2	23.4
Burner tilt (degree)			0		0	0

[a] Proximate analysis [moisture:ash:volatiles:fixed carbon::12:41:23:24], Ultimate analysis [C:H:N:S:O::37:2.26:0.85:0.33:6.53].
[*] Particle size distribution (wt.%) of coal.

Sample	Mass Fraction				Rosin–Rammler Parameters	
	−75 μm	−150 μm to +75 μm	−300 μm to +149 μm	300 μm	Mean Particle Diameter (μm)	Spread Parameter
Subbituminous	0.75	0.166	0.078	0.006	60	1.156

from the open literature. The composition of coal considered in these simulations is provided in Figure 3.6 in Chapter 3. Kinetic parameters for devolatilization and char oxidation as well as gas-phase homogeneous reactions are listed in Table 4.6. The other model parameters used in these simulations are listed in Table 4.7.

The simulation results for a 210-MWe CFD model are discussed below. The predicted results were compared with the few global boiler design parameters obtained from power plants, such as CO_2, H_2O, and O_2 concentrations at the economizer outlet, unburned char in ash, heat transferred to heat exchangers and water walls, FEGT, and the temperature at the inlet of each heat exchanger and at the economizer outlet.

TABLE 4.6

Kinetic Parameters for Devolatilization, Char Oxidation, and Gas-Phase Homogeneous Reactions

Devolatilization (Sheng et al., 2004)		Char Oxidation (Sheng et al., 2004)		Volatile Combustion (Guo et al., 2003)		CO Oxidation (Kim et al., 2000)	
A_v (1/s)	E_v (J/kmol)	A_c (kg/m²sPa)	E_c (J/kmol)	A_{vol} (m³/kmol s)	E_{vol} (J/kmol)	A_{CO} (m³/kmol s)	E_{CO} (J/kmol)
2×10^5	6.7×10^7	0.0053	8.37×10^7	2.56×10^{11}	1.081×10^8	8.83×10^{14}	9.98×10^7

TABLE 4.7

Model Parameters Used for Simulation of Base Case

Parameter	Value	Ref.
Particle emissivity	0.9	Backreedy et al. (2006)
Particle scattering factor (f_p)	0.6	
Wall emissivity	0.6	Gupta (2011)
Swelling factor (S_w)	1	
Heat fraction (f_{heat})	1	Boyd and Kent (1986)
Particle density (ρ_p, kg/m³)	1,400	
Particle heat capacity (Cp_p, J/kg K)	1,680	
Number of particle streams tracked	10,240	
Gas absorption coefficient	WSGGM	

4.2.3.1 Gas Flow

A typical simulated flow field in the Y plane (y = 6.5 m) is shown in Figure 4.20(a) in the form of vector plots. Four distinguished high-gas-velocity jets (20 to 25 m/s) are observed in the burner section of the furnace, representing the combusting mixture of fuel air and coal. It can be seen that most of the gas has upward movement, and a small fraction of gas flows downward toward the ash hopper. The flow moves upward toward the platen

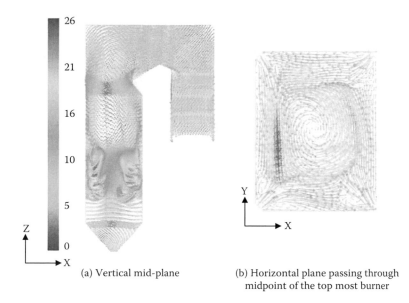

(a) Vertical mid-plane   (b) Horizontal plane passing through
                                    midpoint of the top most burner

FIGURE 4.20 (See Color Insert.)
Simulated gas flow field (velocity shown in m/s): (a) vertical mid-plane and (b) horizontal plane passing through the midpoint of the topmost burner.

superheater and bends around the nose of the furnace to enter into the cross-over pass. From the crossover pass, the flow bends toward the outlet of the boiler. The flow field on the horizontal plane cutting one row of the burners is shown in Figure 4.20(b). It can be seen that the four jets coming from the burners form a strong counter-clockwise circulatory flow. The gas jet, when it enters the furnace, is pushed toward the wall by another jet that is coming from the adjacent wall. This structure is repeated at the four corners that form the rotating fireball at the center. Tangential burners generate rotational flow in the furnace. The swirling nature of the flow continues until the nose section of the furnace is reached. This can be observed in Figure 4.20(a) where, in the upper part of the furnace, the gas has a higher velocity near the walls than that near the center of the furnace. The swirling nature of the flow starts breaking down due to the presence of the suspended platen superheater.

Part of the flow moves toward the front side wall of the boiler, which is still rotational in nature, and then leads to horizontal entry into the platen superheater (SH). While passing through the heat exchangers that are porous volumes, the local velocity of the flue gas increases due to restrictions offered by the heat exchangers. The flow in the crossover pass and rear pass does not show the swirling nature. The gas flow field was also inspected by simulating trajectories of massless particles. The path length distribution and residence time distribution estimated from these simulated trajectories are shown in Figure 4.21(a) and (b), respectively. The mean residence time and traveling length of the flue gas within the boiler was estimated at 11.2 s (standard deviation = 4.5 s) and 73.8 m (standard deviation = 18.6 m), respectively.

4.2.3.2 Particle Trajectories

Trajectories of injected coal particles were also simulated and (colored by z velocity of particles) are shown in Figure 4.22. Because injected coal particles are rather small and the flow is very dilute, most of the particles follow the gas-phase motion and leave the solution domain (boiler) with flue gases. The trajectories show complicated 3D flow characteristics that promote the mixing of the air and coal particles. The initial simulated results show that only ~2% of the ash in the fed coal was recovered as bottom ash (i.e., from the bottom outlet of the furnace) and the remaining ~98% ash was carried away by flue gas as fly ash.

The flue gas and coal particles injected from the lower burners initially circulate in the bottom of the furnace and ash hopper, and eventually travel up through the high-temperature and swirling-flow region (so-called "fireball") formed in the central region of the furnace (Figure 4.22(a)). The flue gas and coal particles from the higher burners pass directly upward from the fireball region (Figure 4.22(b)). The simulated results indicate that nearly 86% of the char and 100% of the volatile material were released in the burner section itself. Of the remaining 14% char, 10% was oxidized in the section above the

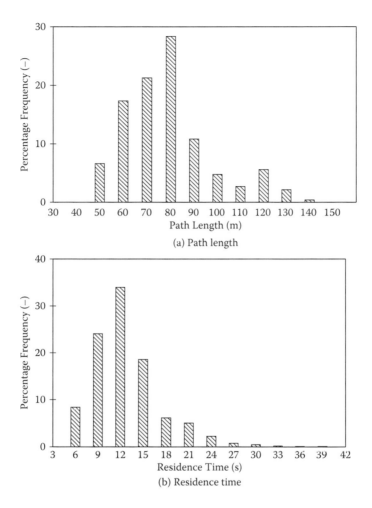

FIGURE 4.21
Simulated distribution of flow path lengths and residence times at the boiler exit.

burner and 4% was oxidized in the bottom section of the burner. The residence times of the flue gas and coal particles injected from the higher burners are shorter in comparison with the flue gas and coal particles injected from the lower burners. The mean residence time of the particles coming out with flue gas is approximately 11 s (standard deviation of 5 s), and that falls down as bottom ash has a mean residence time of around 8 s (standard deviation of 13 s).

4.2.3.3 Temperature Distribution

Simulated values of temperature distribution on a typical vertical plane (plane Y, $y = 6.5$ m and plane Z, $z = 25$ m) are shown in Figure 4.23. The

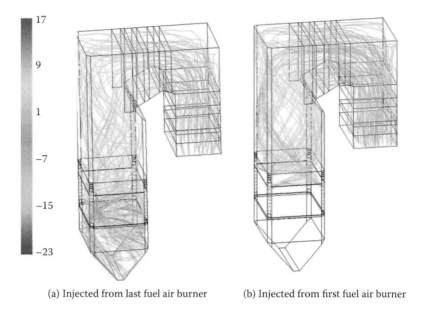

(a) Injected from last fuel air burner   (b) Injected from first fuel air burner

FIGURE 4.22 (See Color Insert.)
Simulated trajectories of coal particles colored by z velocity (m/s) of the particle.

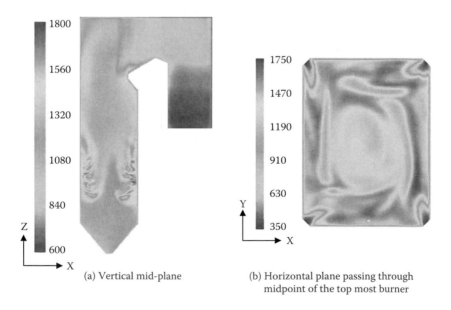

(a) Vertical mid-plane   (b) Horizontal plane passing through
                                 midpoint of the top most burner

FIGURE 4.23 (See Color Insert.)
Typical temperature (K) distribution within a PC fired boiler.

burner section is distinguished by a hot zone in the Y plane. Higher local temperatures (1,500K to 1,800K) are observed in the zone where major combustion reactions are taking place. The temperature distribution on a constant Z plane passing through the fuel air burner port at 25 m is shown in Figure 4.23(b). The cold air jets (350K to 400K) can be observed near the inlets at the corner. The jets get heated due to the heat liberated by combustion reactions along the path. The FEGT ($z = 41$ m) was found to be approximately 1,327K. As the flue gas flows from the furnace exit to the boiler exit, the temperature gradually decreases due to the heat transfer from the flue gas to the furnace walls, reheaters, superheaters, and economizer. The estimated heat flux to the water walls in the burner section was ~121 kW/m² and 90 kW/m² in the sections above and below the burner zone, respectively. The estimated gas-side heat transfer coefficient to the water walls at the burner zone was ~130 W/m²K and around 150 W/m²K for the sections above and below the burner zone, respectively. The average flue gas temperature at the boiler exit (after lower economizer) was 665K.

4.2.3.4 Species Profile

The predicted concentration distributions of oxygen (O₂) and carbon dioxide (CO₂) over a typical vertical plane are shown in Figures 4.24 and 4.25, respectively.

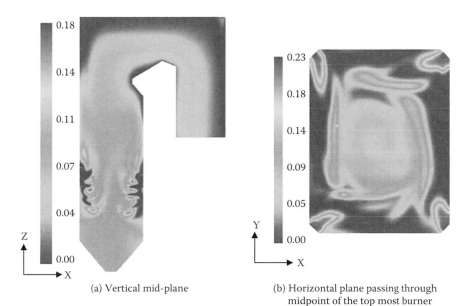

(a) Vertical mid-plane   (b) Horizontal plane passing through midpoint of the top most burner

FIGURE 4.24 (See Color Insert.)
Typical distribution of O₂ mass fraction within a PC fired boiler: (a) vertical mid-plane and (b) horizontal plane passing through the midpoint of the topmost burner.

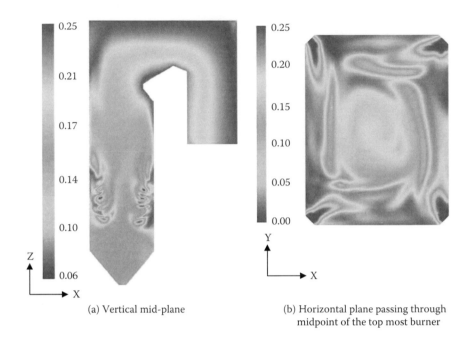

(a) Vertical mid-plane   (b) Horizontal plane passing through
                                          midpoint of the top most burner

FIGURE 4.25 (See Color Insert.)
Typical distribution of CO_2 mass fraction within a PC boiler.

It can be seen that the O_2 concentration is high near the burner tip and rapidly decreases as volatile material and char react with O_2 (Figure 4.24(b)). An opposite trend to this was observed for the CO_2 concentration distribution (Figure 4.25(b)). The value of CO_2 increases from zero at the burner tip to nearly 24% to 25% along the length of the burner jet (Figure 4.25(b)). The O_2 mass was ~4 (mole%) and 3.4 (mole%) at the furnace and boiler exit, respectively. Combustion performance in the boiler is quantified by evaluating unburned char. The model presented here indicates that nearly 95% of the char is reacted in the boiler (see Table 4.8).

4.2.3.5 Heat Transfer to Heat Exchangers

Heat generated due to coal combustion reactions is transferred to the water wall and the suspended superheater and reheater. The predicted heat transferred to internal heat exchangers is listed in Table 4.8, which shows that nearly 40% of total heat transferred is absorbed by the water wall, and the remainder is transferred to the heat exchanger tube bundles. The furnace section of the boiler is primarily radiation dominated; convective heat transfer plays a small role in the overall heat transfer. Quantification of the same is

TABLE 4.8

Summary of Simulated Results for a Considered 210-MWe PC Fired Boiler

(a) Heat Transferred to Heat Exchangers			
	Total Heat	**Relative Contributions to Heat Transfer**	
Heat Exchangers	**Transferred (%)**	**Convective (%)**	**Radiative (%)**
Water wall	39	11	89
Platen SH	22	34	66
Front RH	13	40	59
Rear RH	5.5	45	57
Final SH	3.5	59	41
LTSH	8	37	63
Upper ECO	6	50	49
Lower ECO	3	45	56

(b) Unburned Char in Ash	
Mass flow rate of fly ash (kg/s)	15.52
Mass flow rate of bottom ash (kg/s)	0.3
Total unburned char (%)	5

Note: Total heat transferred is 540 MW.

given in Table 4.8, which shows that nearly 89% of the total energy absorbed by the water wall is via radiative heat transfer.

4.2.3.6 Characteristics of the Crossover Pass

Tangentially fired PC boilers can be operated using a wide variety of coals. In this type of firing system, the fireball formed within the furnace delivers thermal energy to each of the furnace walls. A feature unique to tangentially fired boilers is the ability to regulate furnace heat absorption for steam temperature control, which is done by tilting the fuel and air nozzle assemblies up or down. Superheater (SH) and reheater (RH) temperatures can thus be controlled with minimal plant heat rate impact. Global furnace aerodynamics provides effective furnace volume utilization for higher heat absorption and lower bulk gas temperatures compared to wall-fired boilers. Complete combustion is assured by the combination of maximum residence time and vortex turbulence. However, rotational flow of the tilted tangential firing system affects the thermal load distribution in the convection horizontal gas passage (crossover pass), which increases the thermal load deviation. The development of high-capacity utility boilers, operating at high temperatures and pressures, results in an increased thermal load deviation of the boiler in the lateral direction with horizontal gas passage. Despite its many advantages, this thermal load deviation is inherent in the tangentially fired system and cannot be eliminated completely. The extreme steam temperature deviation experienced in the SH and RH of a utility boiler can seriously influence its economic and safe operation. This temperature deviation is one of the root causes of boiler tube failure (BTF), which causes about 40% of the forced power station outages (Xu et al., 2000). The steam temperature deviation is mainly due to the thermal load deviation in the lateral direction of the SH and RH. This variation is difficult to measure *in situ* using direct experimental techniques. Significant efforts have been expended to predict BTF and determine the mechanisms responsible for BTF by utility boiler companies, boiler manufacturers, and academic researchers (Yin et al., 2002; Dooley and Chang, 2000; Chen, 1997; Xu, 1994; Liu, 1993; Abbott et al., 1992; Yang, 1989, 1991; and Bian, 1987). The developed CFD model was used to quantify the extent of temperature deviation in the crossover pass section. The results are discussed below.

The simulated velocity and temperature field in a crossover path (at a horizontal plane passing through $z = 47$ m) are shown in Figure 4.26. The flow from the furnace section enters the platen superheater from the front wall, and the vector plot shown in Figure 4.26(a) indicates the tendency of flow going toward the right side wall. This is natural for a boiler with counterclockwise firing.

Figure 4.27 shows the velocity plots on the isolines drawn at various (heights) Z locations ($z = 47$ m, 49 m, 51 m) at a particular X distance: (a) x = 8.9 m (before FRH), (b) 11.2 m (before RRH), and (c) 14.2 m (before Final SH) from the front wall. When the flow passes through the platen SH, the flow is guided (due to the presence of the tube bundle) and starts aligning between the right and left walls. However, the effect of imbalance in flow is carried forward to the reheater. The high-velocity zone can be observed in Figure 4.27(b) at the right side wall. When flow comes out of the reheater and goes to the Final SH, the temperature deviation between left and right well diminishes.(see Figure 4.27(c)).

The effect of this mass imbalance leads to higher temperatures on the right side wall and comparatively cooler temperatures on the left side wall. The temperature distribution over an isoplane at height $Z = 47$ m is shown in Figure 4.26. Similar to velocity, the temperatures were plotted at various isolines and are shown in Figure 4.28. These results clearly show a comparatively hotter gas contact with right side wall tubes than left side wall. The quantification of the imbalance of flow was achieved by plotting the temperature deviation along any X isolines. The two temperature values at nearly the same location from the right and left walls were compared, and the difference between right side wall and left side wall temperatures was termed a "temperature deviation." The calculated values of this temperature deviation for various X isolines are shown in Figure 4.29. The effect was prominently observed at the entry of the Front RH and Rear RH, which shows the temperature deviation as high as >150K (maximum ~166K) at height $Z = 47$ m (Figures 4.29(a) and (b)). As we move up toward the roof of the boiler, this

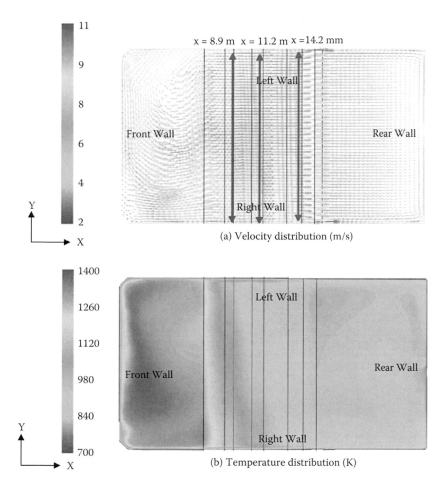

(a) Velocity distribution (m/s)

(b) Temperature distribution (K)

FIGURE 4.26 (See Color Insert.)
Simulated results at crossover pass. (z = 47 m).

deviation decreases. When the flow reaches the entry of the Final SH, it was observed that the deviation decreased to <50K from the right side wall. The simulated value of maximum temperature deviation shows good agreement with the literature value of 150K for 210 MWe (Yin et al., 2002).

4.2.4 Influence of Operating Parameters on Boiler Performance

CFD simulations of 210-MWe PC fired boilers as discussed in the previous section have led to various insights into the behavior of complex processes occurring within the boiler. The model was able capture various key features such as temperature and species profiles, unburned char in ash, heat transfer, and crossover pass characteristics of the boiler. The model may

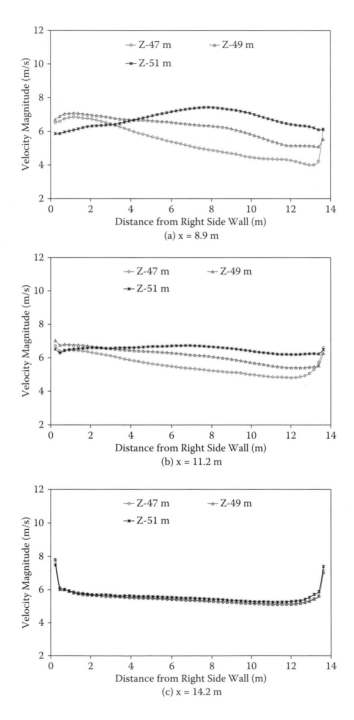

FIGURE 4.27
Profiles of simulated gas velocity (m/s) at crossover pass.

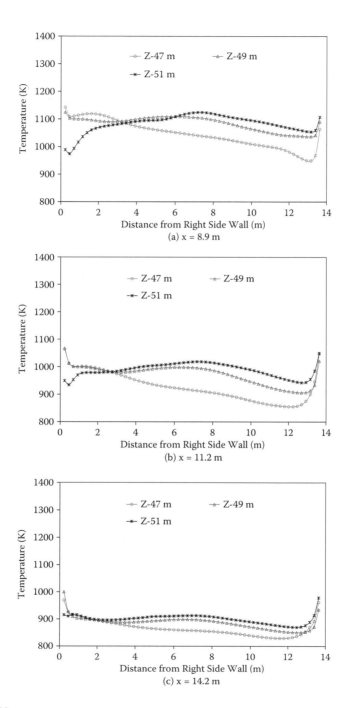

FIGURE 4.28
Profiles of simulated temperatures at crossover pass.

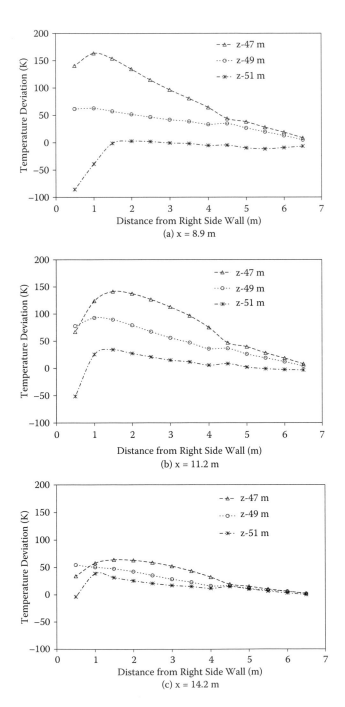

FIGURE 4.29
Simulated profiles of temperature deviation.

be refined further using systematic calibration and validation studies. The methodologies for these are discussed in Chapter 6. Here we will try to illustrate additional applications of the CFD model to understand the influence of various operating parameters on boiler performance.

The overall performance of a boiler depends on many factors, including operating conditions such as local oxygen concentration, coal quality and its properties, burner positions, thermal load, etc. The quantity of coal and air can be monitored before feeding to the boiler. The next monitoring point is the furnace exit gas temperature (FEGT, at height $z = 41$ m in the furnace for the considered case), which is a major indicator of boiler performance and reliability. Management of the FEGT is linked to the optimum controls to be exercised in operating parameters. Undesirable conditions, such as increased slagging/fouling/corrosion rates of the water wall and heat exchanger tubes, occur if there is a deviation in the desired FEGT value. Higher FEGT values will also enhance creep damage and loss of efficiency.

In this illustrative example, sensitivity studies were performed with respect to excess air, coal feed rate, burner tilt, and coal blends. The simulation results are discussed here.

4.2.4.1 Excess Air (i.e., Fuel/Air Ratio)

The optimum excess air, or fuel/air ratio, fed to the boiler is an important factor as it affects unreacted char and pollutant formation. The excess air also absorbs part of the combustion energy to attain the combustion temperature. To illustrate the influence of excess air, the magnitude of excess air was varied from 5% to 20% over the base case discussed previously. The operating conditions are listed in Table 4.5.

The effect of excess air on the temperature distribution along the height of the furnace is shown Figure 4.30. The results show that the cross-sectional average temperature across the furnace height increases by 50K to 100K when excess air changed from 20% to 5%. The effect can be observed even more prominently in the bottom section of the furnace (below the burner zone). The FEGT temperature changed from 1,327K to 1,338K when excess air changed from 20% to 5%.

The observed effect on the oxygen concentration was more profound; see Figure 4.31, where the cross-section average mass fraction of O_2 is shown along the height of the furnace. At the furnace exit, the O_2 mass fraction changed from 0.04791 to 0.0267 when the excess air changed from 20% to 5%. The O_2 mass fraction at the plane $Z = 15$ m at the top of the hopper changed from 0.095 to 0.047 when the excess air changed from 20% to 5%. The reduction in O_2 concentration affects the CO level in the furnace, and it was observed that the CO level at the FEGT increased from 1,448 to 2,753 ppm (Figure 4.32). The excess air affects the combustion efficiency of the boiler, and it was observed that the total unburned char (UBC) in the ash increased from 5% to 8.5% when excess air was reduced from 20% to 5%

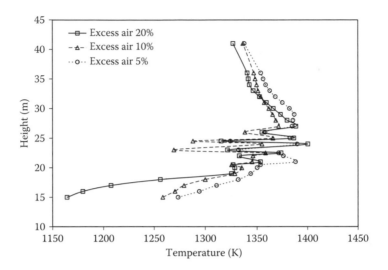

FIGURE 4.30
Influence of excess air on cross-sectional averaged profiles of temperature.

(Figure 4.32). Please note that the exact quantitative influence of excess air on unburned char will be significantly influenced by the coal characteristics and corresponding kinetics of char combustion. The results here are for illustration purposes, and appropriate care should be taken to select suitable char combustion kinetics to reliably simulate the influence of excess air on unburned char.

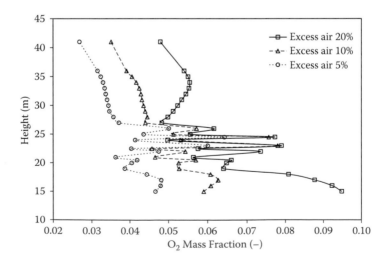

FIGURE 4.31
Influence of excess air on cross-sectional averaged profiles of O_2 mass fraction.

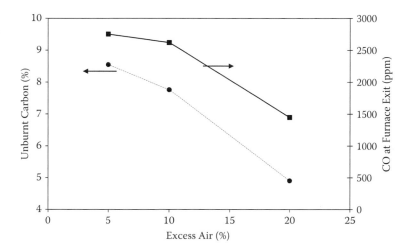

FIGURE 4.32
Influence of excess air on CO concentration (ppm) at furnace exit and unburned char at the
boiler exit.

4.2.4.2 Burner Tilt

Aerodynamics are crucial in the performance of PC fired boilers and can be
drastically influenced by the position and tilt of the burner jets. Burner tilting
results in synchronizing the movement of air nozzles in the vertical direction
(upward/downward) at an angle θ generally less than ±20° (Figure 4.33). The
position of the flame in the furnace can be changed to adjust the superheat
and reheat temperatures. This is useful in changing the combustion conditions for different types of fuel and the heat release rate in the burner area,
and in decreasing slagging and overheat temperature. It should be noted that
there is the danger of creating a serious ash slag in the ash hopper when the
nozzles tilt downward. It is therefore important to quantitatively characterize the influence of burner tilt on the performance of PC fired boilers. The
influence of burner tilt on boiler performance was systematically quantified
in terms of the overall heat transferred to different sections of the water wall
in the furnace and coal burnout for the case of the typical 210-MWe PC fired
boiler considered here.

The effect of burner tilt was simulated by changing the inlet flow velocity conditions such that the fuel air, coal particles, and auxiliary air were
injected at an angle of +20°, +10° in the upward direction, and –10°, –20° in the
downward direction with reference to the horizontal position. The change in
burner tilt shifts the combustion zone in an upward or downward direction,
affecting overall flow characteristics of the boiler. Simulated velocity fields
(as a vector plot at the Y plane, $y = 6.5$ m) for burner tilts +20°, 0°, and 20°
are shown in Figure 4.34. It can be seen that flow also is tilted toward the upper
section or hopper section as per the positive or negative tilt of the burners.

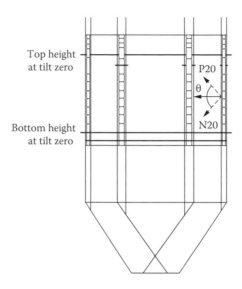

FIGURE 4.33
Schematic showing burner tilt.

Simulated temperature fields (at plane of $y = 6.5$ m) are shown in Figure 4.35. The results clearly show that the combustion reaction zone tilts as per the burner, and therefore heat transfer to water wall characteristics of the furnace are affected. The effect on the heat flux distribution to the water walls above and below the burner section is shown in Figure 4.36. The results show that the heat transfer to the water wall section above the burner (height from

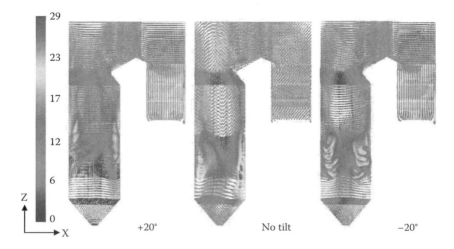

FIGURE 4.34 (See Color Insert.)
Influence of burner tilt on simulated velocity field (m/s) at vertical mid-plane.

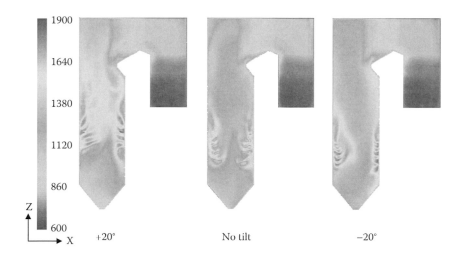

FIGURE 4.35 (See Color Insert.)
Influence of burner tilt on simulated temperature distribution (K) at vertical mid-plane.

$z = 27$ m to 52 m) increases from 87 to 101 MW, and for the section below the burner (height from $z = 9$ m to 20 m) it decreases from 53 to 23 MW when the burner tilt changes from $-20°$ to $+20°$, respectively.

The influence of burner tilt on unburned char (UBC) is shown in Figure 4.37. It was observed that for the base case of zero tilt, the UBC was ~5%. As discussed above, the burner tilt changes the overall dynamics of gas and solid flow, temperature, and species profile in the boiler. It has a

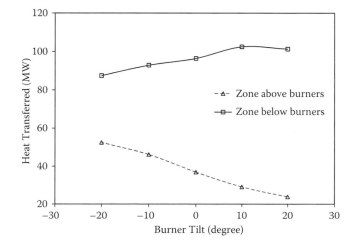

FIGURE 4.36
Influence of burner tilt on heat transferred to water wall.

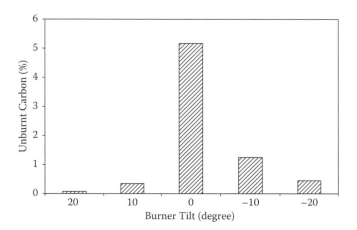

FIGURE 4.37
Influence of burner tilt on total unburned char (UBC) at boiler exit.

significant impact on the residence time of particles in the combustion zone and also the local history of the temperature and oxygen that the particles experience while moving through the furnace. It was observed that the total UBC decreases when the burner tilt is shifted either up or down (Figure 4.37). The results for the +20° tilt showed the lowest UBC of any tilt.

The plane average temperature (T_{plot}) and O_2 concentration (O_{2plot}) profiles along the height of the furnace when the burner tilt was shifted from 0° to ±20° are shown in Figure 4.38 and Figure 4.39, respectively. The combustion or hot zone was identified using these plots of temperature and oxygen mass fractions. The lower coordinate of the combustion zone was marked at a point at which the average plane temperature decreases and the O_2 mass fraction increases sharply. The lower bound for the 0° tilt case was found to be 19 m. A similar approach was useful in the upper section, where the average furnace temperature starts decreasing; it was marked as the upper bound of the reaction/hot zone and observed to be at 27 m. For 0° tilt, the reaction/hot zone was found in the burner section where an array of physical burners (FA and AA) as injecting coal and air into the furnace. Following a similar approach, the lower and upper coordinates of the combustion zones were identified for different burner tilts. The movement of the combustion zone with respect to burner tilt is shown in Figure 4.40.

The influence of burner tilt on the movement of the upper and lower boundaries of the furnace zone may be related to the burner tilt as follows:

- The shift of top plane of the hot zone:

$$Z_T = a_T \, \theta + b_T \tag{4.57}$$

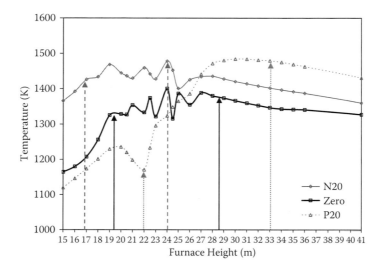

FIGURE 4.38
Influence of burner tilt on cross-sectional averaged profiles of temperature (K).
N20: –20° tilt   zero: no tilt   P20: +20° tilt

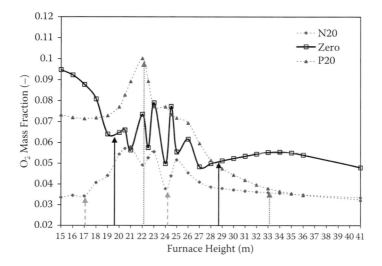

FIGURE 4.39
Influence of burner tilt on cross-sectional averaged profiles of O_2 mass fraction.
N20: –20° tilt   zero: no tilt   P20: +20° tilt

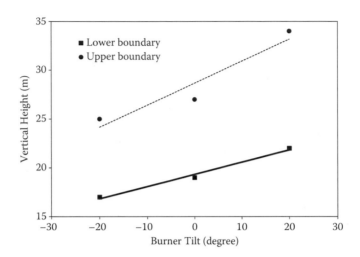

FIGURE 4.40
Influence of burner tilt on movement of upper and lower boundaries of the combustion zone.

- The shift of bottom plane of the hot zone:

$$Z_B = a_B\,\theta + b_B \qquad\qquad (4.58)$$

where Z_T and Z_B are the locations of the top and bottom faces of the fur-nace zone, respectively; θ is the burner tilt angle, in degrees (generally $-20°$ to $+20°$); and a and b are corresponding coefficients, which can be obtained

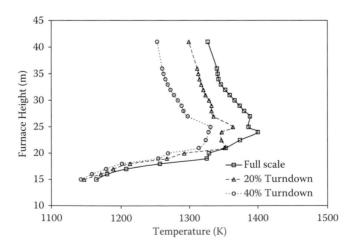

FIGURE 4.41
Influence of turndown on profiles of cross-sectional averaged temperature.

using the simulated results. In the illustrative case considered here, these coefficients were obtained as $a_T = 0.225$, $a_B = 0.125$, and $b_T = 29$, $b_B = 20$.

Such correlations will be useful in developing lower-order models (reactor network models) of PC fired boilers. Development of such models is discussed in Chapter 5.

4.2.4.3 Effect of Boiler Load

The influence of boiler load reduction from normal operating conditions of 210 MWe has been numerically examined. The load was decreased by reducing the coal flow rate (and corresponding air flow rates) by 20% first and then by 40%, distributing evenly to each burner in the operation. The operating conditions are listed in Table 4.5.

The average furnace exit temperature (FEGT) changed from 1,327K (standard operating condition) to 1,252K for the 40% turndown condition (Figure 4.41). Compared with the full load, the cross-sectional average temperature for 40% turndown decreases by nearly 100K. The turndown decreases the gas and particle velocities in the boiler, which increases the residence time of the particles. The mean residence times of the particles for full load and 20% and 40% turndown were found to be 11, 13, and 19 s, respectively. The corresponding extent of char combustion was found to be 95%, 97%, and 97.7%. Figure 4.42 shows the influence of turndown on the heat transferred to the water wall and platen superheater. The total heat transfer to the water wall in the furnace section decreases from 181 to 141 MW; and for platen superheater, the heat transfer decreased from 120 to 74 MW.

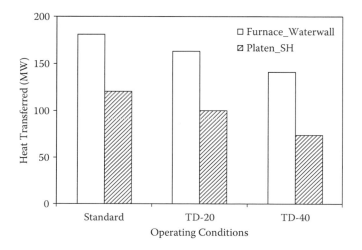

FIGURE 4.42
Influence of turndown on heat transferred to water wall and platen superheater.
TD-20: 20% turndown   TD-40: 40% turndown

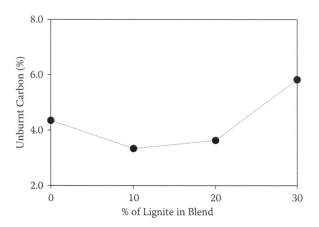

FIGURE 4.43
Influence of share of imported coal in the blend on unburned char.

4.2.4.4 Coal Blends

Blending various types of coal in utility boilers may have a beneficial impact on economics as well as on the environment. It gives better control over pollutant emissions, coal quality, and ash deposition. It is useful in improving combustion behavior, enhancing fuel flexibility, mitigating operational problems (e.g., ash deposition), and reducing fuel costs. Power plants may also passively use blended coals provided by coal suppliers and use coal blending to make a cheaper coal or to modify the properties of a coal with a known problem, such as slagging, sulfur content, volatile content, and/or burnout problems. In past decades, extensive studies were performed to characterize the combustion of coal blends (Sheng et al., 2004; Su, et al., 2003, 2001; Carpenter, 1995). It has been recognized that the properties related to fuel composition (e.g., proximate and ultimate analysis data, heating value, etc.) remain additive after blending, while many characteristics related to combustion exhibit non-additive (i.e., synergistic) behavior. For example, ignition, flame stability, slagging, and fouling cannot be predicted by additivity. Previous studies showed that the combustion behavior of a coal blend is more complex than that of a single coal, and its effects on boiler performance are imperfectly understood.

Experimental approaches have been employed to assess the combustion performance of pulverized coal blends fired in boilers based on bench-scale (Milenkova et al., 2003; Rubiera et al., 2002; Peralta et al., 2002, 2001; Arenillas et al., 2002; Su et al., 2001; Artos, 1993; Beeley et al., 2000; Haas et al., 2001), pilot-scale (Ikeda et al., 2003; Su et al., 2001; Beeley et al., 2000; Maier et al., 1994), and full-scale (Helle et al., 2003; Beeley et al., 2000) data. From the experimental data, some empirical indices, using parameters such as volatile matter content, fuel ratio, and maceral composition, were also derived to empirically predict the ignitability, flame stability, and combustion of coal blends. However, in bench-scale experiments, it was difficult to reproduce the combustion of coal blends prevailing in a practical PC furnace. Investigation of the combustion characteristics of blends via experimental methods is typically expensive and can be labor intensive. There is also a degree of uncertainty involved in extrapolating pilot-scale data directly to full-scale boiler scale. The conclusions are often difficult to extrapolate to unknown coals and blends. The empirical indices, validated by a limited number of experimental data, hence may not be very reliable (Sheng et al., 2004).

It may be possible to use CFD models to understand and simulate the combustion behavior of blends of different rank coal and of high and low ash content coals before actually firing them in a boiler. Recently, significant efforts have been made on the numerical simulation of coal blends and co-firing with biomass (Pranzitelli et al., 2013; Taha et al., 2013; Black et al., 2013; Ghenai and Janajreh, 2010; Shen et al., 2009; Backreedy et al., 2005; Sheng et al., 2004; Arenillas et al., 2002; Beeley et al., 2000). The focus of these studies was to predict the deposition rate of ash particles on the surface of heat exchangers due to slagging and fouling when biomass was used as a co-fuel. Beeley et al. (2000) modeled the combustion of coal blends in a pilot-scale test furnace based on two separate and distinct coal streams with inputs of individual component properties. Arenillas et al. (2002) evaluated the application of CFD to model the combustion of binary coal blends in a bench-scale drop-tube furnace (of internal diameter 20 mm and length 1,420 mm) to predict NO_x emissions and char burnout. In their simulation, one mixture fraction/PDF (probability distribution function) approach was used to model the combustion process, and the blend was represented only as a single coal with properties obtained by weighted averaging relevant properties of the component coals, without adequate description of the chemical and physical interactions between components. Consequently, the non-additivity, particularly from widely different rank coals, was not captured. Sheng et al. (2004) have simulated PC combustion in a pilot-scale furnace (150-kW down fired furnace), and the predictions were compared against the measurements of ignition, burnout, and NO_x emission. Backreedy et al. (2005) demonstrated with experimental validation that the computational models could be applied successfully to both an entrained flow reactor (of internal diameter 40 mm and length 2,000 mm) and an industrial furnace (350 MWe) operated with coal blends. Shen et al. (2009) have simulated pulverized coal combustion in a blast furnace and successfully validated the coal burnout with experimental data. These are encouraging developments over recent decades where CFD models were used to simulate pulverized coal combustion in furnaces.

To illustrate the influence of blends, the CFD model was used to simulate the considered 210-MWe tangentially fired boiler with coal blends of subbituminous coal (high ash, medium volatile) and lignite (low ash, high volatile). The properties of the coal are listed in Table 4.9. The kinetic parameters required for coal combustion were adopted from literature and are

TABLE 4.9

Coal Properties for Blends Considered Here for Illustration

Subbituminous				Lignite			
Proximate		Ultimate		Proximate		Ultimate	
Components	Wt. (%)	Components	Wt. (%)	Components	Wt. (%)	Components	Wt. (%)
Fixed carbon	26.9	C	37.2	Fixed carbon	44.1	C	62.3
Volatiles	21.7	H	2.4	Volatiles	41.3	H	3.8
Ash	47.5	N	0.67	Ash	4.5	N	1.18
Moisture	3.9	S	0.15	Moisture	10.1	S	1.15
		O (by diff.)	8.18			O (by diff.)	17.0
		HHV (kcal/kg)	3260			HHV (kcal/kg)	5840

Kinetic Parameters				
* Refer to Table 4.6 for kinetics of devolatilization and char combustion of subbituminous coal and kinetics of gas-phase reactions	Devolatilization (Zhang et al., 1991)		Char Oxidation (Visona et al., 1999)	
	A_v	E_v	A_c	E_c
	(1/s)	(J/kmol)	(kg/m²sPa)	(J/kmol)
	1.34×10^5	7.41×10^7	0.0042	7.55×10^7

also listed in Table 4.9. The considered operating conditions are shown in Table 4.10.

CFD simulations were performed to predict the burnout behavior of these blends when fired in a full-scale 210-MWe boiler. Simulated results of the UBC for different fuel blends are shown in Figure 4.43. It can be seen that the simulations do not show any systematic trend in UBC with an increase in the share of imported coal (up to 30% of the total blend). The simulated values of UBC were found to be in the range of 3.5% to 5.5%.

TABLE 4.10

Operating Conditions Used in Simulations for Illustrating Influence of Blends

Case	A	B	C	D
Blends of subbituminous and lignite coal (%)	100–0	90–10	80–20	70–30
Fuel air (kg/s); (Temperature 350K)	74.304	74.10	73.945	73.81
Auxiliary air (kg/s); (Temperature 550K)	146.839	146.44	146.129	145.87
Coal flow rate (kg/s); (Temperature 350K)				
Subbituminous	38.05	32.12	26.9	22.25
Lignite (imported)	0	3.56	6.9	9.5

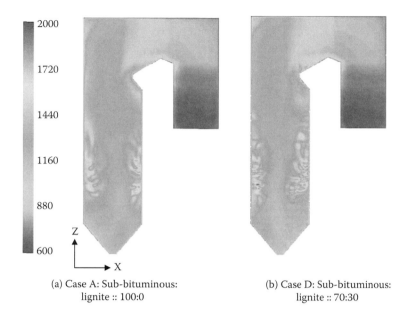

(a) Case A: Sub-bituminous:
lignite :: 100:0

(b) Case D: Sub-bituminous:
lignite :: 70:30

FIGURE 4.44 (See Color Insert.)
Influence of coal blends on simulated temperature (K) distributions at vertical mid-plane:
(a) Case A: Sub-bituminous: lignite :: 100:0, and (b) Case D: Sub-bituminous: lignite :: 70:30.

Simulated temperature distributions on two planes (y = 6.5 m and z = 21 m, FA burner) of the PC boiler are shown in Figures 4.44 and 4.45, respectively, for the A and D blends (and operating conditions) listed in Table 4.10. The simulated temperature distributions were found to be quite different from those for the single coal. This may be because of the different composition as well as different devolatilization and char combustion rates for each type of coal. The maximum flame temperature in the furnace cross-section was found to increase from ~1,850K to 2,112K (see Figure 4.45). This can be expected as the imported coal has a large volatile content (41.3%) that burns faster as compared to char, thus leading to more localized heat generation.

4.3 Summary and Conclusions

Computational fluid dynamics (CFD)-based models offer the possibility of gaining deeper insight into processes occurring in PC fired boilers. Judicious development and use of CFD models may allow one to establish qualitative and quantitative relationships between the performance of a PC fired boiler and various design and operating parameters. The CFD model allows one

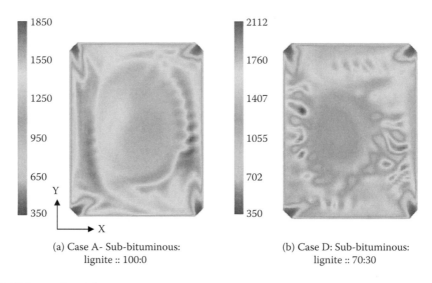

(a) Case A- Sub-bituminous:
lignite :: 100:0

(b) Case D: Sub-bituminous:
lignite :: 70:30

FIGURE 4.45 (See Color Insert.)
Influence of coal blends on simulated temperature (K) distributions at horizontal plane passing through the midpoint of the bottommost burner: (a) Case A: Sub-bituminous: lignite :: 100:0, and (b) Case D: Sub-bituminous: lignite :: 70:30.

to carry out computational experiments to gain a better understanding by simulating the influence of parameters (such as configuration of the boiler; number, location, and design of burners [tilt and swirl at burners]; locations and configurations of internal heat exchangers; coal characteristics [such as quality [coal rank], sulfur and nitrogen content, particle size distribution, composition, and reactivity and its feed rate]; extent of excess air; possibility of air ingress; imbalance in temperatures of steam drums; heat transfer effectiveness, etc.) on boiler performance.

Formulation of model equations for representing flow processes occurring in PC fired boilers was discussed in detail. Flow processes in a PC fired boiler involve turbulent gas-solid flow with homogeneous and heterogeneous chemical reactions along with heat transfer (including radiation). Key issues in modeling these processes and various options available for modeling these were discussed. The model equations for the following choices were presented:

- Turbulence: two-equation turbulence models (the standard k–ε model)
- Gas-solid flow: the Eulerian–Lagrangian approach, DRW model for simulating turbulent dispersion
- Radiation: P-1 radiation model
- Homogeneous chemical reactions: finite rate/eddy dissipation rate model for gas phase combustion
- Heterogeneous chemical reactions: single-step kinetic devolatilization and kinetic/diffusion controlled char oxidation

Some aspects of NO_x formation and particle deposit formation on boiler internals were also discussed.

An illustration of the application of a developed CFD model to simulate a typical 210-MWe tangentially coal fired boiler is included. Methodology for simplifying PC fired boiler geometry using the porous block approach was presented. A unit cell approach to obtain representative characteristics of such porous blocks was discussed and illustrated. The simulated results of boiler performance at normal operating conditions were then presented. The different characteristics, such as temperature profile, species concentration, flow field, and uneven temperature distribution, in the crossover pass were predicted. The computational model was able to simulate the influence of key operating parameters such as

- *Excess air:* The total unburned char (UBC) in the ash increased from ~5% to ~8.5%, when excess air was reduced from 20% to 5%, and the CO level at the furnace exit increased from 1,448 to 2,753 ppm.
- *Burner tilt:* The effect of burner tilt is quantified in the form of a correlation to predict the shift in the hot combustion zone according to the movement of the burner.
- *Thermal load:* The average furnace exit temperature decreased by almost 75K for a 40% turndown condition compared to the standard operating condition.
- *Coal blends:* The model was able to capture the influence of coal blends on char burnout.

The computational model was also able to capture the temperature deviation at the crossover pass and the heat received by various internal heat exchangers of PC fired boilers, including water walls. The models and methodology discussed in this chapter can be used to simulate industrial PC fired boilers. Application of such CFD models for performance enhancement is discussed in Chapter 6. The simulated results of the CFD model will also provide a sound basis for developing "reactor network models" (RNMs) of PC fired boilers. Formulation of RNMs for PC fired boilers is discussed in Chapter 5.

References

Abbott, R.C., Tiley, R., and Gehl, S. (1992). New software directs the fight against boiler tube failure, *Power Engineering*, 96, 59–62.

Arcilla, A.S., Hauser, J., Eiseman, P.R., and Thompson, J.F. (1991). *Numerical Grid Generation in Computational Fluid Dynamics and Related Fields*, North-Holland, Amsterdam.

Arenillas, A., Backreedy, R.I., Jones, J.M., Pis, J.J., Pourkashanian, M., Rubiera, F., and Williams, A. (2002). Modelling of no formation in the combustion of coal blends, *Fuel*, 81, 627–636.

Artos, V. and Scaroni, A.W. (1993). T.G.A. and drop-tube reactor studies of the combustion of coal blends, *Fuel*, 72, 927–933.

Asotani, T., Yamashita, T., Tominaga, H., Uesugi, Y., Itaya, Y., and Mori, S. (2008). Prediction of ignition behavior in a tangentially fired pulverized coal boiler using CFD, *Fuel*, 87, 482–490.

Auton, T.R. (1983). The Dynamics of Bubbles, Drops and Particles in Motion in Liquids, PhD thesis, University of Cambridge, UK.

Backreedy, R.I., Fletcher, L.M., Ma, L., Pourkashanian, M., and Williams, A. (2006). Modelling pulverized coal combustion using a detailed coal combustion model, *Combust. Sci. and Technol.*, 178, 763–787.

Backreedy, R.I., Jones, J.M., Ma, L., Pourkashanian, M., Williams, A., Arenillas, A., Arias, B., Pis, J.J., and Rubiera, F. (2005). Prediction of unburned carbon and NO_x in a tangentially fired power station using single coals and blends, *Fuel*, 84, 2196–2203.

Badzioch, S. and Hawksley, P.G.W. (1970). Kinetics of thermal decomposition of pulverized coal particles, *Ind. Eng. Chem. Process Design and Development*, 9, 521–530.

Baum, M.M. and Street P.J. (1970). Predicting the combustion behavior of coal particles, *Combust. Sci. and Technol.*, 3(5), 231–243.

Baxter, L.L. (1990). In *Coal Combustion Science: Quarterly Progress Report*, Sandia Report SAND 90-8247 (Ed. D.R. Hardesty), Livermore, CA.

Beeley, T., Cahill, P., Riley, G., Stephenson, P., Lewitt, M., and Whitehouse, M. (2000). The effect of coal blending on combustion performance. DTI Report No. COAL R177, DTI/Pub URN 00/509, London.

Belosevic, S., Sijercic, M., Oka, S., and Tucakovic, D. (2006). Three-dimensional modeling of utility boiler pulverized coal tangentially fired furnace, *Int. J. Heat and Mass Transfer*, 49, 3371–3378.

Belosevic, S., Sijercic, M., Oka, S., and Tucakovic, D. (2008). A numerical study of a utility boiler tangentially-fired furnace under different operating conditions, *Fuel*, 87, 3331–3338.

Berlemont, A., Desjonqueres, P., and Gouesbet, G. (1990). Particle Lagrangian simulation in turbulent flows, *Int. J. Multiphase Flow*, 16, 19–34.

Berlemont, A., Simonin, O., and Sommerfeld, M. (1995). Validation of Interparticle Collision Models Based on Large Eddy Simulations, *AME/JSME Int. Conf. of Gas–Solid Flows*, Hilton Head, SC 13–18 August, Proceedings FED, 228, 359–369.

Bian, Q.L. (1987). The experimental study of the boiler tube failure for the high temperature superheater of 1000 ton/h utility boiler at Wangting Power Station, *Boiler Technology*, 4, 1–9.

Black, S., Szuhánszki, J., Pranzitelli, A., Ma, L., Stanger, P.J., Ingham, D.B., and Pourkashanian, M., (2013). Effects of firing coal and biomass under oxy-fuel conditions in a power plant boiler using CFD modelling, *Fuel*, 113, 78–786.

Boyd, R.K. and Kent, J.H. (1986). Three-dimensional furnace computer modeling. In *21st Symp. (Int.) on Combustion*, The Combustion Institute, pp. 265–274.

Brewster, B.S., Smoot, L.D. Barthelson, S.H., and Thornock, D.E. (1995). Model comparisons with drop tube combustion data for various devolatilization sub models, *Energy & Fuels*, 9, 870–879.

Carpenter, A.M. (1995). Coal blending for power stations, IEA Coal Research, London.

Chen, P.P. and Crowe, C.T. (1984). On the Monte Carlo method for modeling particle dispersion in turbulence. In *ASME Symp. Gas–Solid Flows*, p. 37.

Chen, Z.H. (1997). The calculation and analysis of the flow distribution and the thermal variation of the single-phase fluid flowing through the manifold systems in boilers, *J. University of Shanghai for Sci. and Technol.*, 19, 160–178.

Coelho, P.J. (2007). Numerical simulation of the interaction between turbulence and radiation in reactive flows, *Progr. Energy and Combustion Sci.*, 33(4), 311–383.

Davidson, R.M. (1994). *Nitrogen in Coal,* IEA Coal Perspectives, January 1994.

Dean, A.M. and Bozzelli, J.W. (2000). Combustion chemistry of nitrogen, *Gas-Phase Combustion Chem.*, 125–342.

Degereji, M.U., Ma, L., Ingham, D.B., Pourkashanian, M., and Williams, A. (2012). Numerical assessment of coals/blends slagging potential in pulverized coal boilers, *Fuel*, 102, 345–353.

Diez, L.I., Cortes, C., and Pallares, J. (2008). Numerical investigation of NO_x emissions from a tangentially-fired boiler under conventional and overfire air operation, *Fuel*, 87, 1259–1269.

Dooley, B. and Chang, P.S. (2000). The current state of boiler tube failures in fossil plants, *Power Plant Chem.*, 2(4), 197–203.

Eaton, A.M., Smoot, L.D., Hill, S.C., and Eatough, C.N. (1999). Components, formulations, solutions, evaluation, and application of comprehensive combustion models, *Progr. Energy and Combustion Sci.*, 25, 387–436.

Fan, J.R., Jin, J., Liang, X.H., Chen, L.H., and Cen, K.F. (1998). Modeling of coal combustion and NO_x formation in a W-shaped boiler furnace, *Chem. Eng. J.*, 71, 233–242.

Fan, J.R., Zha, X.D., and Cen, K.F. (2001). Study on coal combustion characteristics in a w-shaped boiler furnace, *Fuel*, 80, 373–381.

Fan, J., Sun, P., Zha, X., and Cen, K. (1999). Modeling of combustion process in 600 MW utility boiler using comprehensive models and its experimental validation, *Energy & Fuels*, 13 (5), 1051–1057.

Fan, J., Xha, X. D., Sun, P., and Cen, K. (2001). Simulation of ash deposit in a pulverized coal fired boiler, *Fuel*, 80, 645–654.

Fan, J., Qian, L., Ma, Y., Sun, P., and Cen, K. (2001). Computational modeling of pulverized coal combustion processes in tangentially fired furnaces, *Chem. Eng. J.*, 81(1), 261–269.

Fang, Q., Musa, A.A.B., Wei, Y., Luo, Z., and Zhou, H. (2012). Numerical simulation of multi-fuel combustion in a 200 MW tangentially fired utility boiler, *Energy & Fuels*, 26, 313–323.

Field, M.A. (1969). Rate of combustion of size graded fractions of char from a low-rank coal between 1200K and 2000K, *Combustion and Flame*, 13, 237–252.

Fletcher, T.H. and Kerstein, A.R. (1992). Chemical percolation model for devolatilization: 3. Direct use of ^{13}C NMR data to predict effect of coal type, *Energy & Fuels*, 6, 414.

Fox, R.O. (2003). *Computational Models for Turbulent Reacting Flows*, Cambridge University Press, UK.

Frank, M. and Kalmanovitch, D.P. (1988). An effective model of viscosity for ash deposition phenomena. In: Bryers, R.W. and Ovrres, K.S., Editors, *Mineral Matter and Ash Deposition from Coal*. United Engineering Trustees Inc., 85, 101.

Ghenai, C. and Janajreh, I. (2010). CFD analysis of the effects of co-firing biomass with coal, *Energy Conversion and Management*, 51(8), 1694–1701.

Gosman, A.D., Pun, W.M., Ramchal, A.K., Spalding, D.B., and Wolfshtein, M. (1969). *Heat and Mass Transfer in Recirculating Flows*, Academic Press, London.

Gunjal, P.R., Chaudhari, R.V., and Ranade, V.V. (2005a). Computational study of a single phase flow in packed beds of spheres, *AIChE J.*, 51(2), 365–378.

Guo, Y.C. and Chan, C.K. (2000). A multi-fluid model for simulating turbulent gas–particle flow and pulverized coal combustion, *Fuel*, 79(12), 1467–1476.

Gupta, D.F. (2011). Modeling of Coal Fired Boiler, Ph.D. thesis, University of Pune, India.

Haas, J., Tamura, M., and Weber, R. (2001). Characterisation of coal blends for pulverised fuel combustion, *Fuel*, 80(9), 1317–1323.

Hanson, R.K. and Salimian, S. (1984). Survey of Rate Constants in H/N/O Systems. In W. C. Gardiner, Editor, *Combustion Chemistry*, 361.

Hao, Z., Kefa, C., and Ping, S. (2002). Prediction of ash deposition in ash hopper when tilting burners are used, *Fuel Processing Technol.*, 79(2), 181–195.

He, B., Zhu, L., Wang, J., Liu, S., Liu, B., Cui, Y., Wang, L., and Wei, G. (2007). Computational fluid dynamics based retrofits to reheater panel overheating of No. 3 boiler of Dagang Power Plant, *Computers & Fluids*, 36, 435–444.

Helle, S., Gordon, A., Alfaro, G., García, X., and Ulloa, C. (2003). Coal blend combustion: Link between unburnt carbon in fly ashes and maceral composition, *Fuel Processing Technol.*, 80, 209–223.

Hottel, H.C. and Sarofim, A.F. (1967). *Radiative Transfer*, McGraw-Hill, New York.

Howell, J.R. and Siegel, R. (2002). *Thermal Radiation Heat Transfer, 4th edition*, Taylor & Francis, New York

Howell, J.R. (1968). Application of Monte Carlo to heat transfer problems, *Advances in Heat Transfer*, 5, J.P. Harnett and T. Irvine, Eds., Academic Press, San Diego, pp. 1–54.

Hoy, H.R., Roberts, A.G., and Wilkins, D.M. (1965). Behavior of mineral matters in slagging gasification processes, *J. Inst. Gas Eng.*, 5, 444–469.

Huang, L.Y., Norman, J.S., Pourkashanian, M., and Williams, A. (1996). Prediction of ash deposition on superheater tubes from pulverized coal combustion, *Fuel*, 75(3), 271–279.

Hurt, R., Sun, J., and Lunden, M. (1998). A kinetic model of carbon burnout in pulverized coal combustion, *Combustion and Flame*, 113, 181–197.

Ikeda, M., Makino, H., Morinaga, H., Higashiyama, K., and Kozai, Y. (2003). Emission characteristics of NO_x and unburned carbon in fly ash during combustion of blends of bituminous/sub-bituminous coals, *Fuel*, 82, 1851–1857.

Jones, J.M., Patterson, P.M., Pourkashanian, M., Williams, A., Arenillas, A., Rubiera, F., and Pis, J.J. (1999). Modeling NO_x formation in coal particle combustion at high temperature: An investigation of the devolatilization kinetic factors, *Fuel*, 78, 1171–1179.

Jones, W.P. and Whitelaw, J.H. (1982). Calculation methods for turbulent reactive flows: A review, *Combust. Flame*, 48, 1–26.

Kim, Y.J., Lee, J.M., and Kim, S.D. (2000). Modeling of coal gasification in an internally circulating fluidized bed reactor with draught tube, *Fuel*, 79, 69–77.

Kobayashi, H., Howard, J.B., and Sarofim, A.F. (1977). Coal devolatilization at high temperatures, *Proc. Combust. Inst.*, 16, 411–425.

Kohnen, G., Ruger, M., and Sommerfeld, M. (1994). Convergence behavior for numerical calculations by the Eular/Lagrange method for strongly coupled phases. In *ASME Symp. on Numerical Methods in Multiphase Flows*, 185, 191.

Krawczyk, E., Zajemska, M., and Wylecial, T. (2013). The chemical mechanism of SO_x formation and elimination in coal combustion process, *CHEMIK*, 67, 856–862.

Kumar, M. and Sahu, S.G. (2007). Study on the effect of the operating condition on a pulverized coal-fired furnace using computational fluid dynamics commercial code, *Energy & Fuels*, 21(6), 3189–3193.

Launder, B.E., 1978. Heat and mass transport. In Bradshaw, P. (Ed.), *Topics in Applied Physics 12–Turbulence*. Springer, Berlin.

Launder, B.E. and Spalding, D.B. (1972). *Lectures in Mathematical Models of Turbulence*, Academic Press, London, England.

Lee, F.C.C. and Lockwood, F.C. (1998). Modelling ash deposition in pulverized coal-fired applications, *Progr. Energy Combust. Sci.*, 25(2), 117–132.

Li, Z.Q., Wei, F., and Jin, Y. (2003). Numerical simulation of pulverized coal combustion and NO formation, *Chem. Eng. Sci.*, 58, 5161–5171.

Liu, W.J. (1993). The improved measure and cause analysis of partial overheat for 1025 ton/h utility boiler reheater, *Boiler Technol.*, 7, 4–12.

Lockwood, F.C., Rizvi, S.M., and Shah, N.G. (1986). Comparative predictive experience of coal firing. *Proc. Inst. Mech. Eng.*, 200, 79–87.

Lockwood, F.C., Papadopoulos, C., and Abbas, A.S., 1988, Prediction of a comer-fired station combustor, *Combust. Sci. and Technol.*, 58, 5–24.

Lockwood, F.C. and Shah, N.G. (1981). A new radiation solution method for incorporation in general combustion prediction procedures, *Symp. (Int.) on Combustion*, 18, 1405–1414.

Lokare, S.S. (2008). A Mechanistic Investigation of Ash Deposition in Pulverized-Coal and Biomass, Proquest LLC, Michigan, USA.

Lucas, J.A. (2001). Development of the Thermo-mechanical Analysis Technique for Coal Ash and Slag Applications: Project 3.5 Final Report, Volume 18 of research report (Cooperative Research Centre for Black Coal Utilisation, Australia).

Magnussen, B.F. and Hjertager, B.H. (1976). On mathematical models of turbulent combustion with special emphasis on soot formation and combustion. In *16th Symp. (Int.) on Combustion*, The Combustion Institute.

Maier, H., Spliethoff, H., Kicherer, A., Fingerle, A., and Hein, K.R.G. (1994). Effect of coal blending and particle size on NO_x emission and burnout, *Fuel*, 73, 1447–1452.

Milenkova, K.S., Borrego, A.G., Alvarez, D., Xiberta, J., and Menendez, R. (2003). Tracing the origin of unburned carbon in fly ashes from coal blends, *Energy & Fuels*, 17, 1222–1232.

Modest, M.F. (2003). *Radiative Heat Transfer, 2nd edition*, Academic Press, Burlington, MA, USA.

Morsi, S.A. and Alexander, A.J. (1972). An investigation of particle trajectories in two-phase flow systems, *J. Fluid Mech.*, 55(2), 193–208.

Müller, M., Schnell, U., and Scheffknecht, G. (2013). Modelling the fate of sulphur during pulverized coal combustion under conventional and oxy-fuel conditions, *Energy Procedia*, 37, 1377–1388.

Naruse, I., Yamamoto, Y., Itoh, Y., and Ohtake, K. (1996). Fundamental Study on N_2O Formation/Decomposition Characteristics by Means of Low-Temperature Pulverized Coal Combustion. In *26th Symp. (Int'l) on Combustion*, The Combustion Institute, pp. 3213–3221.

Pallares, J., Arauzo, I., Diez, L.I. (2005). Numerical prediction of unburned carbon levels in large pulverized coal utility boilers, *Fuel*, 84, 2364–2371.

Patankar, S.V. (1980). *Numerical Heat Transfer and Fluid Flow,* Hemisphere, Taylor & Francis Group, New York.

Peralta, D., Paterson, N.P., Dugwell, D.R., and Kandiyoti, R. (2001). Coal blend performance during pulverised-fuel combustion: Estimation of relative reactivities by a bomb-calorimeter test, *Fuel,* 80, 1623–1634.

Peralta, D., Paterson, N.P., Dugwell, D.R., and Kandiyoti, R. (2002). Development of a reactivity test for coal-blend combustion: The laboratory-scale suspension-firing reactor, *Energy & Fuels,* 16, 404–411.

Pillai, K.K. (1981). The influence of coal type on devolatilization and combustion in fluidized beds, *J. Inst. Energy,* 54, 142–150.

Qi, H. and Zhao, B. (2013). *Cleaner Combustion and Sustainable World,* Springer Verlag, Berlin, Germany.

Ranade, V.V. (2002). *Computational Flow Modeling for Chemical Reactor Engineering,* Academic Press, London.

Ranade, V.V., Chaudhari, R.V., and Gunjal, P.R. (2011). *Trickle Bed Reactors, Reactor Engineering and Applications,* Elsevier, United Kingdom.

Richards, G.H., Harb, J.N., and Zygarlicke, C.J. (1992). The effect of variation particle-to-particle composition on the formation of ash deposits. In *Proc., Engineering Foundation Conf. on Inorganic Transformations of Ash Deposition in Combustion,* ASME, New York, pp. 713–732.

Rostam-Abadi, M., Khan, L., DeBarr, J.A., Smoot, L.D., Germane G.J., and Eatough, C.N., (1996). *American Chemical Society, Division Fuel Chem.,* 41(3), 1132–1137.

Rubiera, F., Arenillas, A., Arias, B., and Pis, J.J. (2002). Modification of combustion behaviour and no emissions by coal blending, *Fuel Processing Technol.,* 77, 111–117.

Shen, Y.S., Guo, B.Y., Yu, A.B., and Zulli, P. (2009). A three-dimensional numerical study of the combustion of coal blends in blast furnace, *Fuel,* 88(2), 255–263.

Sheng, C., Moghtaderi, B., Gupta, R., and Wall, T.F. (2004). A computational fluid dynamics based study of the combustion characteristics of coal blends in pulverized coal-fired furnace, *Fuel,* 83, 1543–1552.

Shirolkar, J.S., Coimbra, C.F.M., and Mcquay, M.Q. (1996). Fundamental aspects of modeling turbulent particle dispersion in dilute flows, *Progr. Energy and Combustion Sci.,* 22(96), 363–399.

Sommerfeld, M. (1990). Numerical simulation of the particle dispersion in turbulent flow: The importance of particle lift forces and particle/wall collision models, in *Numerical Methods for Multiphase Flows,* 91, ASME, New York.

Sommerfeld, M. (1993). Reviews in Numerical Modeling of Dispersed Two Phase Flows. In *Proc. 5th Int. Symp. on Refined Flow Modeling and Turbulence Measurements,* Paris.

Spalding, D.B. (1970). Mixing and Chemical Reaction in Steady Confined Turbulent Flames. In *13th Symp. on Combustion,* Salt Lake City, UT, p. 649.

Spalding, D.B. (1971). Concentration fluctuations in a round turbulent free jet, *Chem. Eng. Sci.,* 26,95–107.

Speight, J.G. (2013). *The Chemistry and Technology of Coal, third edition,* CRC Press, Taylor & Francis Group, Boca Raton, FL.

Spinti, J.P. and Pershing, D.W. (2003). The fate of char-N at pulverized coal conditions, *Combustion and Flame,* 135, 299–313.

Spitz, N., Saveliev, R., Perelman, M., Korytni, E., Chudnovsky, B., Talanker, A., and Bar-Ziv, E. (2008). Firing a sub-bituminous coal in pulverized coal boilers configured for bituminous coals, *Fuel,* 87(8-9), 1534–1542.

Srinivasachar, S., Helble, J.J., and Boni, A.A. (1990). Mineral behavior during coal combustion 1. Pyrite transformations, *Prog. Energy Combustion Sci.*, 16, 281.

Srinivasachar, S., Helble, J.J., Boni, A.A., Shah, N., Huffman, G.P., and Huggins, F.E. (1990). Mineral behavior during coal combustion 2. Illite transformations, *Prog. Energy Combustion Sci.*, 16, 293.

Stanmore, B.R. and Visiona, S.P. (2000). Prediction of NO_x emissions from a number of coal-fired power station boilers, *Fuel Processing Technol.*, 64, 25–46.

Su, S., Pohl, J.H., and Holcombe, D. (2003). Fouling propensities of blended coals in pulverized coal-fired power station boilers, *Fuel*, 82, 1653–1667.

Su, S., Pohl, J.H., Holcombe, D., and Hart, J.A. (2001). Slagging propensities of blended coals, *Fuel*, 80, 1351–1360.

Su, S., Pohl, J.H., Holcombe, D., and Hart, J.A. (2001). Techniques to determine ignition, flame stability and burnout of blended coals in p.f. power station boilers, *Progr. Energy and Combustion Sci.*, 27, 75–98.

Shih, T.-H., Liou, W.W., Shabbir, A., and Zhu, J. (1995). A new k-ε eddy-viscosity model for high Reynolds number turbulent flows - Model development and validation, *Computers Fluids*, 24(3), 227–238.

Taha, T.J., Stam, A.F., Stam, K., and Brem, G. (2013). CFD modeling of ash deposition for co-combustion of MBM with coal in a tangentially fired utility boiler, *Fuel Processing Technol.*, 114, 126–134.

Thompson, J.F., Warsi, Z.U.A., and Mastin, C.W. (1985). *Numerical Grid Generation*, North-Holland, Amsterdam.

Thomson, D.J. (1987). Criteria for the selection of stochastic models of particle trajectories in turbulent flows, *J. Fluid Mech.*, 180, 529–556.

Tomeczek, J., Palugniok, H., and Ochman, J. (2004). Modelling of deposits formation on heating tubes in pulverized coal boilers. *Fuel*, 83(2), 213–221.

Toor, H.L. (1975). The non-premixed reaction, *Turbulence in Mixing Operations*, Ed. R.S. Bradkey, Academic Press, New York.

Turns, S. R. (2000). *An Introduction to Combustion: Concepts and Applications*, McGraw-Hill International Editions: Mechanical Engineering Series, Singapore

Urbain, G., Bottinga, Y., and Richet, P. 1982. Viscosity of liquid silica, silicates and aluminosilicates, *Geochim. Cosmochim. Acta*, 46, 1061–1072.

Urbain, G., Cambier, F., Deletter, M., and Anseau, M.R. 1981. Viscosity of silicate melts, *Trans. Br. Ceram. Soc.*, 80, 139–141.

Vargas, S., Frandsen, F., and Dam-Johansen, K. (2001). Rheological properties of high-temperature melts of coal ashes and other silicates, *Progr. Energy Combustion Sci.*, 27(3), 237–429.

Viskanta, R. and Menguc, M.P. (1987). Radiation heat transfer in combustion system, *Progr. Energy and Combustion Sci.*, 13, 97–160.

Visuvasam, D., Selvaraj, P., and Sekar, S. (2005). Influence of Coal Properties on Particulate Emission Control in Thermal Power Plants in India. In *Proc. Second Int. Conf. Clean Coal Technologies for Our Future* (CCT 2005). 2005, Sardinia, Italy.

Vuthaluru, R. and Vuthaluru, H.B. (2006). Modelling of a wall fired furnace for different operating conditions using FLUENT, *Fuel Processing Technol.*, 87, 633–639.

Walsh, P.M., Sayre, A.N. Lochden, D.O., Monroc, L.S., Beer, J.M., and Sarofim, A.F. (1990). Deposition of bituminous coal ash on an isolated heat exchanger tube: Effects of coal properties on deposit growth, *Prog. Energy Combustion Science*, 16, 327.

Wang, H. and Harb, J.N. (1997). Modeling of ash deposition in large-scale combustion facilities burning pulverized coal, *Progr. Energy Combustion Sci.*, 23, 267–282.

Williams, A., Pourkashanian, M., and Jones, J.M. (2001). Combustion of pulverised coal and biomass, *Progr. Energy and Combustion Sci.*, 27, 587–610.

Winter, F., Wartha, C., Loffler, G., and Hofbauer, H. (1996). The NO and N_2O Formation Mechanism During Devolatilization and Char Combustion under Fluidized Bed Conditions. In *26th Symp. (Int.) on Combustion*, The Combustion Institute, 3325–3334.

Xu, L.J. (1994). Research and analysis of boiler tubes failure diagnosis for superheater and reheater, *J. Energy Research and Utilization*, 33 (3), 34–38.

Xu, M., Azevedo, J.L.T., and Carvalho, M.G. (2000). Modelling of the combustion process and NO_x emission in a utility boiler, *Fuel*, 79(13), 1611–1619.

Xu, M., Yuan, J., Ding, S., and Cao, H. (1998). Simulation of the gas temperature deviation in large-scale tangential coal fired utility boilers, *Comput. Methods Appl. Mech. Eng.*, 155, 369–380.

Xu, M., Azevedo, J.L.T., and Carvalho, M.G. (2001). Modeling of a front wall fired utility boiler for different operating conditions, *Comput. Meth. Appl. Mech. Eng.*, 190, 3581–3590

Yang, R.G. (1989). BTF cause and diagnosis for the reheater of 1160 ton/h utility boiler at Baoshan Iron and Steel Group Corp., *Boiler Technology*, 8, 1–8.

Yang, R.G. (1991). The improved measure and diagnosis of boiler tube failure for the reheater of 1160 ton/h utility boiler, *Boiler Technol.*, 2, 5–8.

Yin, C., Caillat, S., Harion, J., Baudoin, B., and Perez, E. (2002). Investigation of the flow, combustion, heat-transfer and emissions from a 609 MW utility tangentially fired pulverized-coal. *Fuel*, 81(8), 997–1006.

Zbogar, A., Flemming, J.F., Jensen, P.A., and Glarborg, P. (2005). Heat transfer in ash deposits: A modelling tool-box, *Progress in Energy and Combustion Sci.*, 31(5–6), 371–421.

Zhou, H., Zhou, B., Zhang, H., Li, L., and Cen, K. (2014). Investigation of slagging characteristics in a 300 kW test furnace: Effect of deposition surface temperature, *Ind. Eng. Chem. Res.*, dx.doi.org/10.1021/ie4041516.

Zhou, L.X., Li, L., Li, R.X., and Zhang, J. (2002). Simulation of 3-D gas-particle flows and coal combustion in a tangentially fired furnace using a two-fluid-trajectory model, *Power Technol.*, 125, 226–233.

5

Reactor Network Model (RNM) of a Pulverized Coal Fired Boiler

Computational fluid dynamics (CFD)-based models of pulverized coal (PC) fired boilers require large computational resources and turnaround times. Therefore, such models are generally not appropriate for quick analysis or for on-line optimization (and are also not intended for such). Classical chemical reaction engineering (CRE) approach-based models rely on simplified assumptions about flow and mixing in the reactor, and focus on the solution of mass and energy conservation equations. In these models, because momentum equations are not solved, demands on computational resources are orders of magnitude lower than for CFD models. These models also offer quick turnaround times and therefore have the potential for use with on-line optimization. However, for complex reactors such as PC fired boilers where flow, heat transfer, and chemical reactions are intricately coupled, it is rather difficult to make *a priori* assumptions about the flow and mixing for formulating CRE models.

It is therefore logical to apply detailed flow and mixing knowledge from CFD to formulate CRE models. One approach that is usually used for liquid-phase reactors is to carry out nonreactive CFD simulations and use the simulated results to formulate a CRE model as well as quantify parameters related to flow and mixing in CRE models. The reduced-order CRE models typically use multiple zones or a network of ideal reactors to represent complex systems. The number of such zones or reactors in a network is typically on the order of hundreds as compared to millions of computational cells used in CFD models. The size, location, and mass exchanges among the zones/reactors are usually specified using the simulated results of the CFD model. A more general approach, however, is to integrate fully reactive CFD models directly with computationally efficient, reduced-order reactor network models (RNMs). There may be different levels of integration between CFD and RNM models. In this chapter, development of RNMs for PC fired boilers is discussed. An overall approach toward the development of RNMs is discussed in Section 5.1. An illustration of the application of this approach to develop RNMs for a typical 210-MWe PC fired boiler is then discussed in subsequent sections.

5.1 Approach to Develop Reactor Network Models

Reactor network models (RNMs) represent an actual reactor as several inter-connected, idealized zones (or reactors). These idealized reactors are typically chosen to be completely mixed or tubular reactors. For a realistic description of real reactor behavior, RNMs must be formulated based on prevailing flow patterns and mixing within the reactor. In the early years, chemical engineers used concepts such as residence time distribution (RTD) and state of mixed-ness to account for non-ideal flow and mixing in reactors. Experimental information and intuitive understanding of flow also have been used to formulate RNMs. Mann and co-workers have developed models based on 200 to 400 completely mixed zones or cells with finite exchange between neighboring zones reactors connected in such a way as to represent the actual flow generated by impellers (Mann and Mavros, 1982; Wang and Mann, 1992). In these models, they used experimental measurements of velocity data for prescribing the flow through different zones. They were reasonably successful in simulating reactor performance for fast, mixing-controlled reactions. However, an approach that relies either on RTD data and a qualitative understanding of the flow patterns, or on experimental measurements of flow, to establish an RNM has obvious limitations for general application.

Several significant efforts were made in subsequent years to develop and apply RNMs. The most notable contributions have been made by the groups of Ray (Ray and Wells, 2005 and references cited therein) and Pantelides (Bezzo et al., 2003 and references cited therein). Both groups have developed and evaluated various approaches for formulating RNMs using the simulated results of the CFD model. Pantelides and co-workers developed a general approach for determining flow quantities from nonreactive CFD simulations and applying them for formulating RNMs. Ray and co-workers proposed various methodologies to use spatial segregation of the temperature or composition for formulating RNMs. Mujumdar and Ranade (2009) used CFD simulations to manually formulate RNMs for simulating rotary cement kilns. Gupta (2011) and Gupta and Ranade (2008) have similarly developed RNMs from detailed CFD simulations of PC fired boilers. Concepts similar to RNMs were also used for simulating complex heat transfer; see, for example, the zone model by Hottel and Sarofim (1967), and the Monte Carlo model by Gosman (1973) and Howell (1968). These models solve for the heat transfer using intuitive information about the gas flow pattern (Lowe et al., 1975).

Recent advances in computing resources, numerical techniques, and our understanding of turbulent, reacting multiphase flows have made CFD simulations of complex systems such as PC fired boilers not only possible, but also increasingly common. Illustration of the development and application of CFD models for PC fired boilers was provided in Chapter 4. Automated formulation of RNMs from CFD simulations is very appealing because it can establish strong and sound relationships between CFD models and RNMs.

However, based on our experience using RNMs for a variety of different complex reactors—for example, *para*-xylene oxidation in stirred reactors, hydrogenation in bubble columns, hydro-desulfurization in trickle beds, rotary cement kilns, and PC fired boilers (Ranade et al., 2011; Ranade and Mujumdar, 2009; Ranade, 2002)—here we prefer not to use automated formulation of RNMs from the CFD simulations despite the natural appeal of that option. Instead, we have used an approach that relies on loose coupling between CFD and RNMs. It may be noted that CFD simulation of a complex system such as a PC fired boiler is complicated and may have several uncertainties associated with various submodels used within the overall CFD model. These uncertainties associated with the various submodels are not always quantifiable and may be quite uneven across various submodels. Rigorous error analysis is virtually impossible for simulations of industrial PC fired boilers with the usual allocated resources for CFD simulations. In light of this, our experience indicates that manual intervention at the formulation stage of RNMs from CFD simulations is useful. Eventually, automated formulation of RNMs from CFD simulations may be practically feasible and better than the approach proposed here. However, at present and in the near future, we believe that manual intervention in the formulation of RNMs is optimal from the application-to-practice point of view. Various approaches to formulate RNMs and the present approach are shown schematically in Figure 5.1.

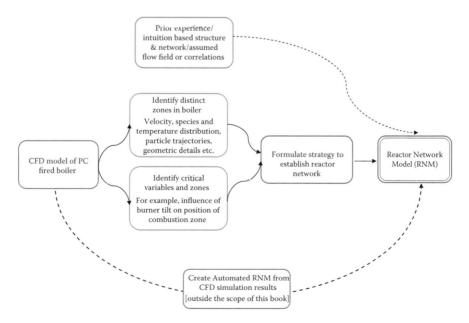

FIGURE 5.1
Overall approach to develop reactor network model (RNM).

The methodology for formulating RNM from CFD results is discussed in Section 5.2.

5.2 Formulation of Reactor Network Models from CFD Simulations

As mentioned previously, RNMs represent PC fired boilers using a network of ideal reactors/zones. Detailed information on flow, heat transfer, and reactions (distribution of species concentration) within the PC fired boiler obtained using CFD simulations is used for formulating a network of reactors to represent the PC fired boiler. Mass and energy conservation equations are then developed for the established reactor network. Appropriate solution methodology and simulation strategies are then developed to allow quick simulations using RNMs. The developed RNM can then be used for exploring a wider parameter space or for on-line optimization and process control. In this section, the formulation of reactors/zones and connections among these reactors (network) are discussed.

5.2.1 Formulation of Reactors/Zones in a PC Fired Boiler

In this step, the considered solution domain of a PC fired boiler is divided into a number of ideal reactors (completely mixed reactor or plug flow reactor). It is also essential to understand the various parameters that may change the number, size, and locations of these identified reactors/zones. For example, in PC fired boilers, multiple burners are used. The reactors (number, size, and location) representing PC fired boilers may vary, depending on how many burners are active (and their locations). The tilt of the burners can also be varied during boiler operation. Variation in burner tilt also influences the number, size, and locations of reactors included in the RNM. It is therefore necessary to account for this while formulating the RNM. The overall steps in identifying zones/reactors within PC fired boilers can therefore be written as follows:

- Divide the PC fired boiler region into different zones.
- Develop criteria to identify distinct ideal zones/reactors within each of these regions.
- Use heuristics and/or CFD results for defining the location, shape, and size of zones/reactors.
- Identify variables such as coal flow rate, excess air, burner tilt, etc. and capture their influence on identified locations and sizes of zones/reactors.

These steps are discussed in this section. Use of CFD results to connect identified zones for establishing a network is discussed in Section 5.2.2.

The PC fired boiler has three major regions:

1. Furnace
2. Crossover pass (CP)
3. Second pass (SP)

These regions are shown in Figure 1.1 in Chapter 1 and Figure 5.6 in this chapter.

The furnace region has a complex flow structure, especially in the burner zone (see Figures 4.20, 4.22, and 4.23 in Chapter 4). Air and (or) a coal jet from the burner are injected at a predefined angle into the furnace and form a rotating flow pattern in the cross-sectional plane of the furnace; this is called a *fireball*. Coal, air, and combustion products travel along the length of each burner jet and discharge the same into the fireball. This mechanism creates a swirling flow inside the furnace that then travels upward to the exit of the furnace. As most of the reactions take place in the burner zone, it can also be called a *combustion zone*. The furnace region is first characterized, and different zones within this region are identified as a primary step toward developing an RNM. The combustion zone is an important section of the boiler where the coal and air are injected; it is shown schematically in Figure 5.2.

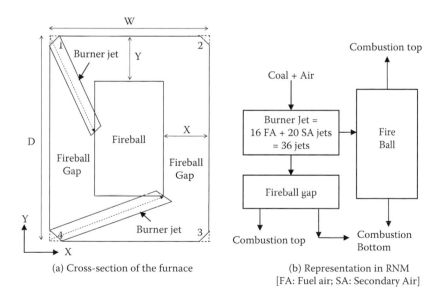

(a) Cross-section of the furnace

(b) Representation in RNM
[FA: Fuel air; SA: Secondary Air]

FIGURE 5.2
Schematic of combustion zone: (a) cross-section of the furnace and (b) representation in an RNM (FA = fuel air, SA = secondary air).

The combustion zone has significant variations in concentration and temperature, and therefore may have to be further divided into distinct zones. To illustrate this, a typical 210-MWe PC fired boiler was considered. There are sixteen fuel-air burners used in this boiler. Each of the fuel-air burners is sandwiched between two secondary air inlets. In many PC fired boilers, over fire air (OFA) inlets may also be used for manipulating the formation of pollutants such as NOx. However, for the particular case considered in this illustrative example, OFA inlets were not used. Therefore, there were a total of thirty-six jets (sixteen FA [fuel air] + twenty SA [secondary air] jets) injected from the four corners of the furnace. Based on the simulated flow, temperature, and species field in the combustion zone (see Chapter 4), the combustion zone can be subdivided into burner jet, fireball, and fireball gap zones (see Figure 5.2). From CFD simulations it is evident that all these burner jets join in a central zone called the fireball. In principle, it is possible to consider and define thirty-six burner jet zones to account for all these jets. Instead, more practical approaches would be to consider four burner jet zones corresponding to each corner of the furnace, or to merge all these thirty-six jets into a single burner jet zone. In this illustrative case study, the latter option was selected. If the interest is to also understand the influence of the possible maldistribution of gas and solid flow at each corner, then each corner jet may have to be modeled separately or even each jet may have to be modeled separately.

After identifying the burner jet zone, the next step is to establish the dimensions (width, depth, and height for rectangular shape assumption) of this zone. A simplest approximation would be to consider the dimension as that of the burner slit (refer to Figure 5.2). Another approach would be to analyze the contour plots of the gas velocity, species concentration, and temperature for a plane passing through any burner that is injecting coal plus air (refer Figures 4.20, 4.22, 4.24, and 4.25 in Chapter 4). Based on an analysis of these data, it is also possible to estimate the width of the burner jet. For the illustrative example here, the burner jet dimension was assumed to be the same as the burner slit. All the jets were represented by a single jet, and therefore the cross-sectional area of the single burner jet is the sum of the cross-sectional area of all thirty-six burner ports/slits. The length of the burner jet depends on the dimensions of the fireball (refer to Figure 5.2). The dimensions and shape of the fireball must be defined. The fireball can either be assumed as rectangular (brick shape in three dimensions [3D]) or circular (cylindrical shape in 3D).

For the rectangular fireball, the width and depth must be defined (X and Y as shown in Figure 5.2(a)). Once the values of X and Y are obtained, the height of the fireball can be set equal to the height of the combustion zone. Using the dimensions of the fireball, burner, and combustion zone, the dimensions and volume of the fireball gap can be estimated. The values of X and Y can be obtained from CFD results in a variety of ways, including

- Iso-surfaces of temperature in the furnace region
- Iso-surface of oxygen mass fraction in the furnace region
- Profile of temperature along x and y axes on the plane passing through the center of the fuel burner
- Profile of oxygen mass fraction along x and y axes on the plane passing through the center of the fuel burner
- Profile of temperature deviation along the x and y axes (difference between local temperature and average temperature across the line) on the plane passing through the center of the fuel burner
- Profile of deviation of oxygen mass fraction along x and y axes (difference between local oxygen mass fraction and average oxygen mass fraction across the line) on the plane passing through the center of the fuel burner
- Profile of some combination of deviation of oxygen mass fraction and temperature deviation (e.g., product of these two deviations) along x and y axes on the plane passing through the center of the fuel burner

It may be difficult to decide *a priori* which way is better suited for a given configuration of a PC fired boiler. For the illustrative case considered here, iso-surfaces of temperatures in a typical 210-MWe PC fired boiler are shown in Figure 5.3. It can be seen that depending on the choice of temperature, the size

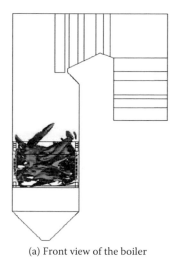
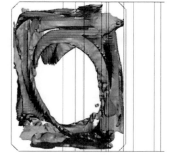

(a) Front view of the boiler   (b) Top view of the furance

FIGURE 5.3 (See Color Insert.)
Iso-surface of temperature in furnace zone (1,550K): (a) front view of the boiler and (b) top view of the furnace.

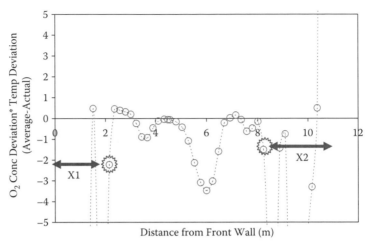

(a) Sample of deviation of O_2 mass fraction X deviation of T on line passing through center of the COMBUSTION zone (parallel to x-axis)

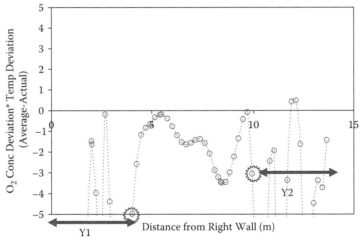

(b) Sample of deviation of O_2 mass fraction X deviation of T on line passing through center of the COMBUSTION zone (parallel to y axis)

FIGURE 5.4
Estimation of the dimensions of a fireball.

of the fireball will change. The shape of the iso-surface also appears to deviate from a rectangular box. As an example of an alternative, the product of the temperature deviation and the deviation in oxygen mass fraction from the average along the x and y axes are shown in Figure 5.4. Once the profile of the desired variable (from the list given above or otherwise) is obtained, the boundaries of the fireball can be identified by defining some criteria. For example, the width or depth of the fireball can be defined as when the deviation values go beyond a certain range. Using the profiles shown in Figure 5.4 and selecting the range of +5 to −5, the boundaries of the fireball can be obtained by identifying points denoted by X1,X2 and Y1,Y2. Such an exercise can be repeated for several iso-lines at constant x and constant y values on the z plane passing through the center of the fuel burner, and average width or depth based on these several pairs of X1,X2, and Y1,Y2 can be estimated. Based on such an analysis, the average values for X and Y were estimated as X = 2.26 m and Y = 3.98 m (cor-responding fireball dimensions therefore are 5.48 m by 6.04 m).

The temperature of the fireball region may be more or less similar to that of the burner jets. However, the O_2 concentration in the fireball will be lower than that in the jets as it is continuously depleting in the jets. Generally, it is observed that the fireball width and depth cannot be captured adequately using only temperature or only O_2 mass fraction profiles. Some combination of temperature and oxygen mass fraction often provides better estimates of fireball dimensions.

It should be noted that as the number of active burners or the burner tilt changes, the location of the combustion zone will also change. The change in active burner positions will simply move the z position of the combus-tion zone accordingly. The position of the combustion zone and therefore the burner jet, fireball, and other connected zones will depend on burner tilt in a complex way. The upward or downward tilt of burners will correspond-ingly move the combustion region, and corresponding fireball zone, in the respective direction. The movement of the fireball position with burner tilt is quantified from the CFD results and is shown for the illustrative case con-sidered here in Figure 5.5.

The position of the combustion zone is adjusted as per the position of the fireball zone. The volume remaining in the combustion zone after subtract-ing the volumes of the burner jet and the fireball is called the fireball gap (Figure 5.2(a)). The walls surrounding the gap zones are the water walls. The heat transfer area for each section of the water wall is obtained from geomet-ric details available in the boiler manual. The fireball gap can be modeled either as four zones (attached to four water walls) or as a single zone. For applying an RNM to the illustrative case considered here, we used the latter option. This completes the process of zoning the combustion zone, which provides the basis and starting point for continuing the process of defining zones for other parts of the PC fired boiler.

CFD results show that the gas flow coming up from the combustion zone is swirling in nature, and hence the top section of the combustion zone is rep-resented by two coaxial zones, and a similar assumption is extended to the zone below the combustion zone as shown in Figure 5.6. The width and depth of the central volume were assumed to be the same as the fireball. The area and volume for zones such as the nose and above that were estimated from the geometric details of the boiler. It should be noted that the dimensions of

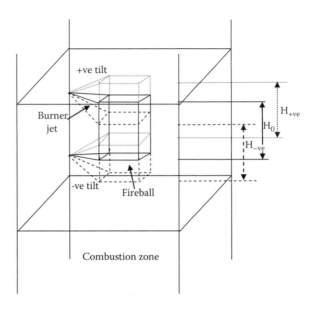

(a) Schematic of the FIREBALL zone at different burner tilts H_0, H_{-ve} and H_{+ve}: FIREBALL position at zero tilt, downward tilt and upward tilt of burner respectively.

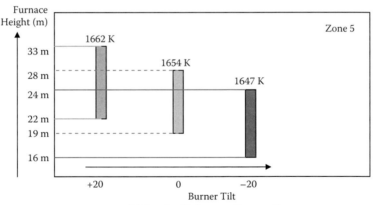

(b) Sample of influence of burner tilt on temperature and location of FIREBALL zone

FIGURE 5.5 (See Color Insert.)
Effect of burner tilt on fireball zone.

zones such as the "burner top" and "burner bottom" depend on the dimensions of the combustion zone. The height of the combustion zone was estimated from the burner tilt correlations obtained from CFD simulations (refer to Chapter 4, Figure 4.40). According to the change in burner tilt, the height of the combustion zone changes as discussed in Chapter 4, and therefore the dimensions of the zones above and below the combustion zone will

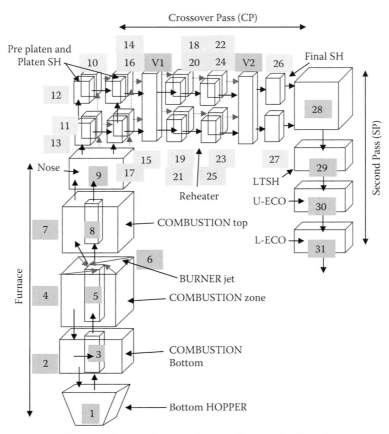

Zone number in yellow color boxes indicates right side wall and
similarly blue color indicates left side wall of the boiler,
Green arrow indicates the burner jets.

FIGURE 5.6 (See Color Insert.)
Reactor network model for tangentially fired PC boiler.

1: HOPPER;
2: COMBUSTION BOTTOM (CB) GAP; 3: COMBUSTION BOTTOM (CB) CORE;
4: FIREBALL GAP; 5: FIREBALL; 6: BURERJET;
7: COMBUSTION TOP (CT) GAP; 8: COMBUSTION TOP (CT) CORE;
9: NOSE; 10 TO 13: PRE PLATEN; 14 TO 17: PLATEN;
V1: VIRTUAL VOLUME; 18 TO 21: FRONT RH; 22 TO 25: REAR RH;
V2: VIRTUAL VOLUME; 26 TO 27: FINAL SH;
28: PASS2TOP; 29; LTSH; 30: U-ECO; 31:L-ECO

change (Figure 5.5). For example, if the burners are tilted 20° upward, then
the combustion zone will move toward the combustion top, and this change
in height of the combustion zone due to burner tilt is estimated from the tilt
correlations discussed in Chapter 4 (Equations (4.67) and (4.68)). These equations provide new height and location of the combustion zone based on how
much the combustion top zone height is reduced and combustion bottom

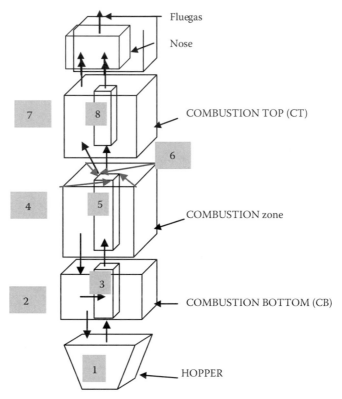

FIGURE 5.7 (See Color Insert.)
Schematic of a reactor network for the bottom section.

1: HOPPER;
2: COMBUSTION BOTTOM (CB) GAP; 3: COMBUSTION BOTTOM (CB) CORE;
4: FIREBALL GAP; 5: FIREBALL; 6: BURERJET;
7: COMBUSTION TOP (CT) GAP; 3: COMBUSTION TOP (CT) CORE

zone height is increased in order to keep the total volume of three zones constant (refer to Figures 5.5 and 5.7). This effect is reversed if burners are tilted in the downward direction.

The crossover pass and second pass zones (platen, reheater, final SH, LTSH, and economizer) have heat exchanger bundles. Radiative and convective heat are transferred from the flue gas to these tube bundles. It is a common observation in tangentially fired PC boilers that there is a higher temperature toward one side of the boiler than on the opposite wall (temperature deviation). This happens mainly because of the uneven flow distribution in the crossover pass. The detailed CFD analysis of such uneven flow distribution and the corresponding temperature deviations were discussed in Chapter 4. To capture the temperature deviation effect in an RNM, the zones in the crossover pass can be divided into four sections: top left, top right, bottom left, and bottom right zones (Figure 5.6). This configuration can be used for the crossover pass up to the reheater section. For the final SH, only the top and bottom sections can be considered because it was observed in CFD simulations that the deviation effect diminishes when flue gas reaches the final SH (Chapter 4, Figure 4.29). The remaining sections of the second pass, such as the LTSH and economizer, can be considered as mixed zones reactors in series.

The reactor network shown in Figure 5.6 can be subdivided further into smaller zones if desired. The zones, such as burner top, burner bottom, reheater, LTSH, etc., can be represented as multiple reactors/zones in series. The reactor network can also be constructed with built-in flexibility to select an appropriate number of reactors/zones. The formulated reactor network can be fine-tuned with the help of CFD simulations. After establishing the reactors/zones, the next step is to establish a connecting network and estimate related quantities for these connections. This is discussed in Section 5.2.2.

5.2.2 Formulation of Reactor Network from the Identified Zones

The network and corresponding flow rates/mass exchanges among the formulated zones should be established using the CFD results. To illustrate this, simulated CFD results for a typical 210-MWe PC fired boiler are used. These results were discussed in Chapter 4.

The simulated CFD results clearly point out that the burner jets feed combusting material to the fireball. The vector plot shown in Figure 4.4 indicates that part of the flow moves down toward the hopper section, and the remaining flow moves toward the top section of the furnace. The flue gas that flows down from the fireball enters the combustion bottom (CB) gap zone. Here it further divides into flow toward the hopper and the CB-core. The flue gas that comes to the hopper then moves upward through the CB-core and then through the fireball gap. The flue gas that comes to the CB-core from the CB-gap also moves upward through fireball gap. Based on this analysis, the reactor network for the bottom section is shown in Figure 5.7.

The distributions of the z velocity at the bottom plane of the combustion zone (below the lower-most operating FA burner) and at the top plane of the hopper zone are shown in Figure 5.8. The mass flow rate of flue gas flowing downward was determined on clip planes of negative z velocity. Based on this, information about the distribution of flow at these locations was obtained.

The burner tilt also affects the flow distribution in the furnace. CFD results obtained at different burner tilts were therefore analyzed (see Figure 5.9). It can be seen that there is a significant influence of burner tilt on the flow distribution in the lower part of the furnace. The influence of burner tilt on the percentage of total flue gas mass flow rate toward the lower part of the furnace may be correlated as

$$F_{BC}\% = a_{BC} \cdot \theta + b_{BC} \tag{5.1}$$

$$F_{TH}(\%) = a_{TH} \cdot \theta + b_{TH} \tag{5.2}$$

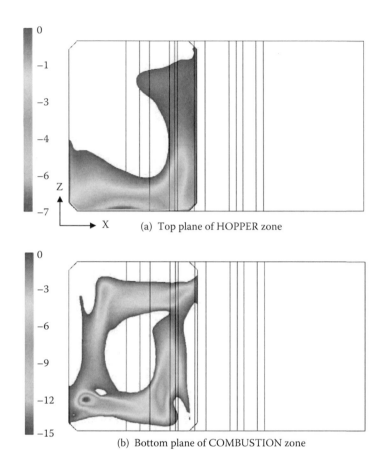

(a) Top plane of HOPPER zone

(b) Bottom plane of COMBUSTION zone

FIGURE 5.8 (See Color Insert.)
Region showing downward flow at two horizontal planes below the combustion zone (contours of z velocity, m/s).

where F is the percentage of total air (mass flow rate) at the inlet of the boiler, and θ is the burner tilt angle (in degrees) with respect to the horizontal axis. The subscripts BC and TH indicate bottom plane of the combustion zone and top plane of the hopper zone, respectively. The coefficients a and b appearing in Equations (5.1) and (5.2) are obtained from the simulated flow field (Figure 5.9). Equations (5.1) and (5.2) can therefore be used to quantify the influence of burner tilt on the flue gas distribution in the zones below the combustion zone.

The flue gas from the combustion zone moves upward toward the nose of the boiler through the combustion top (CT) zone. The structure of the CT is the same as CB, and was split into two parts as CT-core surrounded by CT-gap. The dimensions of the CT-core were assumed to be the same as

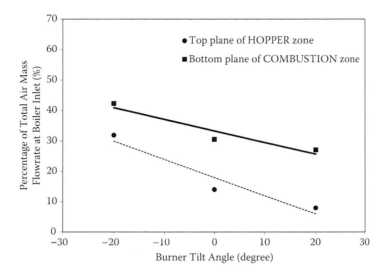

FIGURE 5.9
Influence of burner tilt on mass flow distribution in the lower part of furnace.

the fireball. The flue gas that moves upward splits into these two coaxial zones and for each zone, the mass flow rate was estimated from the flow field obtained using the CFD model. It was observed that around 30% of the total flue gas flow goes to the CT-core. The flow distribution in each of these zones was estimated from CFD simulations. The section above the CT is the nose, and it can be modeled as a single mixed reactor. It collects the mass from the CT-gap and CT-core. The flue gas from the nose moves toward the crossover pass of the boiler. The zone above the nose can be split into two equal volumes: the pre-platen and platen zones. The reactor network of this part of the PC fired boiler is shown in Figure 5.10.

The influence of burner tilt on the flow distribution at the crossover pass was also analyzed based on CFD simulations (see Figure 5.11). It can be seen that there is a negligible influence of burner tilt on the flue gas mass flow rate distribution toward the right side zone (including top and bottom) of the front reheater (front RH). Therefore, for this illustrative example, the RNM model ignored the influence of burner tilt on flow distribution in the crossover pass. It is, however, possible to develop correlations similar to Equations (5.1) and (5.2) to account for the influence of burner tilt.

The platen zone has a platen SH tube bundle, whereas the pre-platen was a void space between the tube bundles and boiler water walls. Part of the flue gas coming from the nose enters this zone. For a typical counter-current fireball observed in a tangential PC fired boiler, it was observed through CFD simulations that in the upper furnace zone, the fluid has a tendency to move toward the right side wall of the boiler. As discussed earlier, this leads

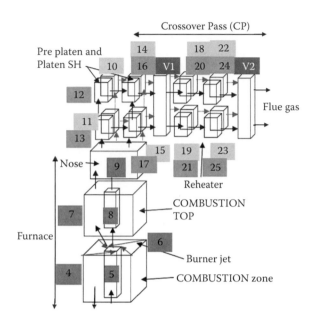

FIGURE 5.10 (See Color Insert.)
Reactor network for fireball-CT-nose section.
4: GAP;
5: FIREBALL;
6: BURERJET;
7: COMBUSTION TOP (CT) GAP;
8: COMBUSTION TOP (CT) CORE;
9: NOSE;
10 TO 13: PRE PLATEN;
14 TO 17: PLATEN

to a temperature deviation in the upper furnace zone and crossover pass zone on the left and right side walls. Zones such as the pre-platen, platen, and front and rear reheaters were split into four subzones as top-left, top-right, bottom-left, and bottom-right. Because the influence of this deviation diminishes after the reheater, the final superheater was modeled as two sub-zones (top and bottom). The mass flow distribution for each of these zones can be estimated from the CFD results in a straightforward way. The flow network from platen onward is shown in Figure 5.12. The fluid from the final superheater passes toward the PASS2TOP, LTSH, UECO, and LECO. Each zone may be further divided into "n" internal reactors if necessary. From the platen to the LECO zones the gas exchanges heat with the water wall/steam wall and heat exchangers. There are two virtual volumes provided that act as a flow mixer/splitter for the next zone. The V1 virtual volume collects the

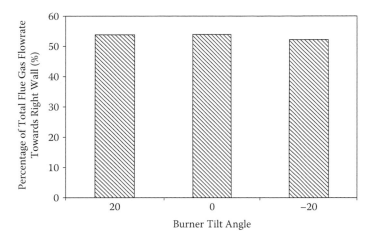

FIGURE 5.11
Influence of burner tilt on mass flow distribution in the crossover pass (front RH).

flow from all four platen SHs (zones 14 to 17) and redistributes the flow into the next four zones of the front reheater (zones 18 to 21) based on the outlet flow fraction assigned to V1. Introduction of such virtual volumes is useful because flow redistribution/realignment occurs due to the resistance offered

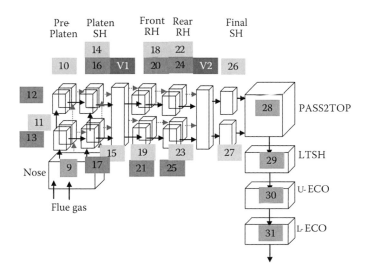

FIGURE 5.12 (See Color Insert.)
Reactor network for platen superheater to economizer section.
9: NOSE; 10 TO 13: PRE PLATEN; 14 TO 17: PLATEN;
V1: VIRTUAL VOLUME; 18 TO 21: FRONT RH; 22 TO 25: REAR RH;
V2: VIRTUAL VOLUME; 26 TO 27: FINAL SH;
28: PASS2TOP; 29: LTSH;30:U-ECO;31:L-ECO

by the tube bundles. Based on the CFD results, appropriate outlet flow fractions can be assigned to these virtual volumes.

Particle motion in PC fired boilers should also be represented adequately in the RNM. Appropriate equations for simulating particle-level processes such as heating, devolatilization, and reaction can be formulated using the Lagrangian framework. However, because flow of the continuous phase is not solved in the RNM, appropriate information regarding continuous-phase velocity must be specified. A critical analysis of simulated particle trajectories from the CFD model was carried out. The motion of particles through the different zones of the RNM was quantified, and an average residence time spent in each of the zones was analyzed. Typical trajectories through zones of the RNM and mean residence time values in different regions of the PC fired boiler are listed in Table 5.1.

Particles enter the domain through burners and leave the domain either via the bottom hopper (bottom ash) or from the LECO zone (fly ash). The mass flow distribution of the particles that are moving down from the combustion zone toward the bottom of the furnace was estimated from the CFD results. A typical fraction of particles moving down toward the bottom of

TABLE 5.1

Motion of Injected Coal Particles through Various Zones of a PC Fired Boiler

Particle Stream 1		Particle Stream 2		Particle Stream 3		Particle Stream 4	
Zone	Residence Time (s)	Zone	Residence Time (s)	Zone	Residence Time (s)	Zone	Residence Time (s)
6	0.5	6	0.5	6	0.5	6	0.5
5	1.0	5	1.0	4	2.0	5	1.0
7	2.0	8	2.0	2	1.0	4	1.0
9	0.7	9	0.7	3	1.0	2	3.0
10	0.1	13	0.3	5	1.0	1	2.0
13	0.2	17	0.5	8	1.0	Total	7.5
17	0.5	19	1.0	9	0.5		
19	1.0	21	1.0	12	0.5		
26	2.0	26	1.0	14	0.5		
28	1.0	28	1.0	18	0.5		
29	0.5	29	0.5	20	0.5		
30	0.5	30	0.5	24	0.5		
31	1.0	31	1.0	28	0.5		
Total	11	Total	11	29	0.5		
				30	0.5		
				31	0.5		
				Total	11.5		

Note: Zone numbers and residence times predicted from CFD simulations of the illustrative case considered here. Please refer to Figure 5.1 for zone locations (1: bottom; 6: burner inlet; 31: flue gas exit).

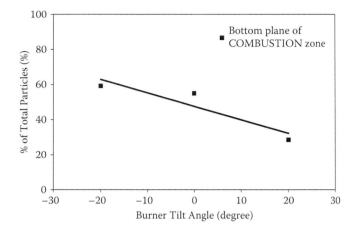

FIGURE 5.13
Influence of burner tilt on downward motion of particles (fraction of total particles moving toward bottom section of the furnace).

the furnace at z = 20 m as a function of burner tilt is shown in Figure 5.13. The simulation results can be used to obtain suitable correlations to estimate the influence of burner tilt on the particle mass flow rate distribution in the bottom section of the furnace.

Information on the distribution of particles and mean residence times in different zones of the RNM is used while simulating particle-level processes. This completes the basic formulation methodology for the RNM; the model equations for each zone are described in Section 5.3. In this section, the reactor network is formulated manually using the CFD results. As discussed previously, eventually an algorithm along the lines proposed by Ray and co-workers or Pantelides and co-workers may be implemented to automate the formulation of the reactor network from the CFD simulations.

5.3 Model Equations and Solution

Each zone in the RNM formulated to simulate a PC fired boiler consists of two phases: the continuous gas phase and the discrete solid phase (coal particles). Mass and energy conservation equations must be developed for both phases over each of these zones/reactors. Particle-level processes can be modeled using the Lagrangian framework. Here, the coal devolatilization and char oxidation reactions were modeled following the practices used in the CFD simulations (Chapter 4). Homogenous gas-phase reactions were modeled using the eddy dissipation concept. Radiative heat transfer was modeled based on the Hottel Zone method (Hottel and Sarofim, 1967). It should be mentioned here that although this section discusses a specific set of model equations corresponding to the illustrative example considered herein, the approach is quite general. The approach can be extended in a straightforward manner to include additional reactions or to replace some of the approximations used while formulating these model equations. The model equations are presented in the following subsections. Each zone is made up of "n" internal reactors, which help to divide each zone into a number of zones/internal reactors. In this current illustration to keep the approach simple, the zones such as the fireball, gap, hopper, pre-platen, platen, RH, and final SH are assumed to be single internal reactors; that is, each zone will have a single well-mixed reactor. Other zones, such as the burner jet, combustion bottom, combustion top, LTSH, and economizer, have the ability to increase the number of internal CSTRs to a user-defined value. Nevertheless, any of the above zones can be further split into a larger number of internal reactors to provide better resolution of parameters such as temperature, species composition, etc.

5.3.1 Continuous Phase

The continuous gas phase was modeled using the Eulerian approach. Each zone (refer to Figure 5.6) will be represented by the superscript k and, as discussed in the above paragraph, each zone can be further split into a number of internal reactors/CSTRs, which is represented by the superscript n. Hence, the following conservation equations are written for each nth reactor of the kth zone.

- Overall mass balance:

$$\frac{d\left(\rho^{k,n} V^{k,n}\right)}{dt} = \sum_{i \neq k} \left(\Phi^{k,k} F_{out}^{i,n}\right) - F_{out}^{k,n} + \sum_{j} S_{j}^{k,n} \qquad (5.3)$$

where n is any internal reactor of the kth zone, i is any zone other than k, and j is any gas component. F is the mass flow rate (kg/s) coming into/going out of the CSTR, and $\Phi^{k,k}$ provides details about which zone's outlet will be connected to the current k^{th} zone as an inlet. Its value varies between 0 and 1 and provides the fraction of total gas flow coming out of any other zone and the same will become an inlet to the current k^{th} zone.

- Component balance: The species conservation equation for the nth internal reactor of kth zone can be written as

$$\frac{d\left(\rho^{k,n} V^{k,n} m_{m}^{k,n}\right)}{dt} = \sum_{i \neq k} m_{m}^{i,n} \Phi^{k,k} F_{out}^{i,n} - m_{m}^{k,n} F_{out}^{k,n} + R_{m}^{k,n} + S_{m}^{k,n} \qquad (5.4)$$

where $\rho^{k,n}$ is the gas density of the kth zone and n^{th} internal CSTR (kg/m³), V is the volume of nth internal CSTR of the kth zone (m³), m_m is the mass fraction of species m, R_m is the net rate of production or consumption of species m by chemical reaction (kg/s), and S_m is the source of species m from the dispersed phase (kg/s).

- Energy balalnce: The enthalpy conservation equation for the nth internal reactor of kth zone can be written as

$$\frac{d\left(\rho^{k,n}V^{k,n}h^{k,n}\right)}{dt} = \sum_{i \neq k} h_{out}^{i,n}\Phi^{k,k}F_{out}^{i,n} - h_{out}^{k,n}F_{out}^{k,n} + S_{gas-rxn}^{k,n} + S_{char}^{k,n} + S_{rad}^{k,n} \quad (5.5)$$

where h is the enthalpy:

$$h^{k,n} = \sum_{j} m_m^{k,n}h_m^{k,n} \quad (5.6)$$

where h_m is the enthalpy of the mth species defined as

$$h_m^{k,n} = h_m^0 + \int_{T_{ref}}^{T^{k,n}} C_{p,m}\,dT \quad (5.7)$$

where h_m^0 is the standard heat of formation of species j (J/kg), $S_{gas-rxn}$ is the source term of heat of chemical reactions (W), S_{rad} is the heat transfer by radiation from all other zones (W), and S_{char} is the source term for discrete-phase char oxidation (W).

The heat released due to chemical reactions is

$$S_{rxn}^{k,n} = \sum_{r} \Delta H \cdot R_r^{k,n} \quad (5.8)$$

The heat of reaction, ΔH, is defined as

$$\Delta H = \sum v_i\, h_{f,i} \quad (5.9)$$

where v_i is stoichiometric coefficient of species i and $h_{f,i}$ is heat of formation of species i. Summation is over all species participating in a reaction. Heat transfer to the water wall and heat exchangers of the each reactor due to convection is

$$S_{Conv-ww}^{k,n} = h_{ww}^{k,n}A_{ww}^{k,n}\left(T_g - T_{ww}\right)^{k,n} + h_{HTX}^{k,n}A_{HTX}^{k,n}\left(T_g - T_{HTX}\right)^{k,n} \quad (5.10)$$

Heat transfer coefficients for convective heat transfer to heat exchangers were estimated using correlations of the Nusselt number (see, for example, Equation (4.55)). The parameters of the correlation were taken from Table 4.4.

The heat transfer coefficient for the water walls was estimated from the simulated (from CFD model) results of the total convective heat transferred to water walls. More details of the modeling of homogeneous gas-phase reactions and radiation are discussed in Sections 5.3.3 and 5.3.4. The surface area of water wall was estimated from the outer diameter, length, and number of the tubes passing through each section of the furnace. Similarly, the surface area of the heat exchanger for each zone was estimated from the characteristic dimensions of the tube bundle present in each zone. Information about the tube outer diameter, tube length, and number of tubes in each tube bundle required for estimation of surface area can be obtained from the boiler manual and engineering drawings, which are generally available from the boiler supplier.

5.3.2 Discrete Phase

The discrete phase was modeled using the Lagrangian approach. To reduce the burden on computational resources, the mass flow rate of the coal at the burner inlet is represented by a small number of finite streams. Information about the motion of these particles from different zones and their mean residence time in different zones was obtained from CFD simulations of the PC fired boiler. Particle trajectories simulated using the CFD model were also used to estimate particle flow rate fractions (of the total coal mass flow rate at the inlet) in each stream. For example, the motions of four such particle streams through different zones of the RNM and their corresponding mean residence times in different zones are listed in Tables 5.1 and 5.2, respectively.

Because the motion and mean residence times of the representative particle streams are obtained from the CFD model, it is not necessary to solve particle momentum equations within the RNM. For each particle stream, mass and energy balance equations are formulated and solved. It is essential to account for particle size distribution while formulating these balance equations. As discussed in Chapter 4, the particle size distribution was described using the Rosin–Rammler equation (see Equation (3.12) in Chapter 3 and Figure 5.14). The considered particle size distribution in this illustrative example is shown in Figure 5.14.

For the illustrative case considered here, the Rosin–Rammler parameters were a mean particle diameter of 60 µm and a spread parameter of 1.156

The mass and energy balance equations for the discrete phase (coal particles) are discussed next.

5.3.2.1 Particle Mass Balance

The change in mass of a particle of size j is the mass loss due to devolatilization, and char oxidation may be represented as

$$\frac{dM_{p,j}}{d\theta} = \left(\frac{dM_{v,j}}{d\theta} + \frac{dM_{c,j}}{d\theta} \right) \qquad (5.11)$$

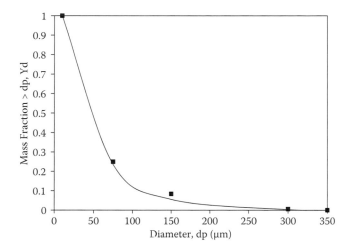

FIGURE 5.14
Particle size distribution data (shown by symbols) represented using the Rosin–Rammler distribution (mean particle diameter is 60 µm, and the spread parameter is 1.156).

where θ is the time spent by the particle in each zone, and it was assumed that all particles spend the same amount of time in the reactor, irrespective of their particle size or density. $M_{c,j}$ is the mass of fixed carbon present at any time in the particle of size j, and $M_{v,j}$ is the mass of volatiles present at any time in the particle of size j. The mean time spent by particles in each zone was estimated from the CFD simulations by tracking the particles at the inlet and outlet of each concerned zone (see Table 5.1). The devolatilization and combustion were modeled as follows:

- *Devolatilization: The coal devolatilization rate for a particle of size* j *can be written as (Badzioch and Hawksley, 1970)*

$$\frac{dM_{v,j}}{d\theta} = -A_v\, e^{(-E_v/RT_{P,j})} M_{v,j} \qquad (5.12)$$

where $M_{v,j}$ is the mass of the volatile in particle of size j at any time (kg), E_v is the activation energy for devolatilization (J/kmol), A_v is the preexponential factor for devolatilization (1/s), and T_p is the temperature of the particle (K).

In the devolatilization phase, the particle diameter was assumed to remain constant and the particle density was allowed to decrease to account for the reduction in particle mass due to the devolatilization.

- *Surface reaction-char oxidation:* The char combustion rate was calculated based on the assumption that the char is oxidized to CO by the following reaction:

$$C_{(s)} + 0.5\, O_{2\,(g)} \rightarrow CO_{(g)} \quad \Delta H^\circ = -110 \; kJ/mol$$

The rate of char oxidation can be written as (Baum and Street, 1970; Field, 1969):

$$\frac{dM_{c,j}}{d\theta} = - A_{p,j} \frac{K_c K_{d,j}}{K_c + K_{d,j}} P_{O2} \tag{5.13}$$

The kinetic rate constant (K_c) for the char oxidation reaction is

$$K_c = A_c \, e^{(-E_c/RT_P)} \tag{5.14}$$

where A_c is a preexponential factor for char combustion (kg/m²-s-Pa), and E_c is the activation energy for char combustion (J/kmol).

The bulk gas-phase diffusion coefficient (K_d) for the oxidant (Field, 1969) can be given as

$$K_{d,j} = \frac{5 \times 10^{-10}}{d_{p,j}} \left(\frac{T_g + T_{p,j}}{2} \right)^{0.75} \tag{5.15}$$

The change in density of the coal particle of particle size j can be written as follows (Smith, 1971):

$$\rho_{p,j} = \rho_{po,j} \left(U_j \right)^{\beta} \tag{5.16}$$

$$d_{p,j} = d_{po,i} \left(U_j \right)^{\alpha} \tag{5.17}$$

where $3\alpha + \beta = 1$

U is the overall unburned fraction of coal and can be written as

$$U = \sum_j U_j = \sum_j \frac{M_{v,j} + M_{c,j} + M_{w,j} + M_{A,j}}{M_{v0,j} + M_{c0,j} + M_{w0,j} + M_{A0,j}} \tag{5.18}$$

where $M_{v,j}, M_{c,j}, M_{w,j},$ and $M_{A,j}$ are the mass of volatile, char, moisture, and ash present in the particle at any time, and the subscript "0" indicates the mass of volatile, char, moisture, and ash initially present in the coal particle. In this illustrative case, the size of the particle during char oxidation was assumed to remain unchanged. Therefore, values of α and β were set to zero and one, respectively, in Equations (5.16) and (5.17). The particle density varies as the fixed char present in the particle is oxidized.

5.3.2.2 Particle Energy Balance

An energy balance over a dispersed phase can be written as

$$M_{p,j}Cp_p\frac{dT_{p,j}}{d\theta} = \left(f_{heat}Q_{char,j}\right) + Q_{rad,j} + Q_{conv,j} \tag{5.19}$$

where Cp_p, H_{rxn}, Q_{rad}, and Q_{conv} are the particle specific heat, heat of char oxidation reaction, and radiative and convective heat transfer, respectively. The f_{heat} is the fraction of heat liberated during char oxidation ($Q_{char,j}$), which is absorbed by the particle.

$$Q_{char,j} = \frac{dM_{c,j}}{d\theta}H_{char-rxn} \tag{5.20}$$

The radiative heat transfer can be written as

$$Q_{rad,j} = \varepsilon_P\sigma A_{p,j}(T_g^4 - T_{p,j}^4) \tag{5.21}$$

where ε_P is the emissivity of the particle, σ is the Stefan–Boltzmann constant = 5.67×10^{-8} W/m²-K⁴, $A_{p,j}$ is the particle surface area of size j (m²), and T_g is the gas-phase temperature (K).

The convective heat transfer can be written as

$$Q_{conv,j} = h_{c,j}A_{p,j}(T_g - T_{p,j}) \tag{5.22}$$

The heat transfer coefficient h_c was evaluated using the correlation of Ranz and Marshall (1952) as

$$\frac{h_{c,j}d_{p,j}}{k_g} = 2 + 0.6(\text{Re}_p)^{0.5}(\text{Pr})^{0.33} \tag{5.23}$$

where h is the heat transfer coefficient for gas and solid flow (W/m²-K), k_g is the conductivity of the gas (W/m-K), Re_p is the particle Reynolds number, and Pr is the Prandtl number.

The particle-level processes represented in the balance equations discussed above will provide source or sink terms to the continuous phase. For example, the rate of consumption of oxygen by a heterogeneous reaction with coal particles will provide a sink term to the continuous-phase oxygen balance equation. Therefore, the mass source term from the discrete phase can be written as

$$S_{k,n} = \int_0^{\tau_{k,n}} (R_v + R_c)\, d\theta \tag{5.24}$$

where R_v and R_c are the rate of devolatilization and char oxidation, and $\tau_{k,n}$ is the residence time of the particle in any n^{th} internal reactor of the kth zone.

$$R_v = \sum_{j=1}^{N} N_{p,j} \frac{dM_{v,j}}{d\theta}, \quad R_c = \sum_{j=1}^{N} N_{p,j} \frac{dM_{c,j}}{d\theta} \tag{5.25}$$

where N is the number of size classes used for representing the particle size distribution, and $N_{p,j}$ is the total number of particles of size class j. In an illustrated example, four particle streams were tracked, and the mass distribution to each stream was assumed based on a simulated distribution of the fly ash and bottom ash generated in the boiler. The particle trajectory of each stream is provided in Table 5.1, where particle streams 1, 2, and 3 are for fly ash and stream 4 is for bottom ash. The particle size distributions for each stream were assumed to be identical and were obtained from the Rossin–Rammler function.

5.3.3 Homogenous Gas-Phase Reactions

Homogeneous gas-phase reactions can be modeled using the species transport model discussed in Chapter 4. The effective rate of mixing can be estimated using the values of turbulent kinetic energy (k) and turbulent energy dissipation rates (ε) obtained from CFD simulations. The net source of chemical species j due to reaction is computed as the sum of the rate of reactions over the N_r reactions that the species participate in as

$$R_j = V\, Mw_{,j} \sum_{r=1}^{Nr} R_{j,r} \tag{5.26}$$

The volatile material was represented by single species as $CH_{2.08}O_{0.33}$ based on proximate and ultimate analyses of the coal. The following homogenous gas-phase reactions were considered:

$$CO_{(g)} + 0.5\, O_{2(g)} \rightarrow CO_{2(g)} \quad \Delta H^\circ = -283\ kJ/mol$$

$$CH_{2.08}O_{0.38\,(g)} + 1.33\, O_{2\,(g)} \rightarrow CO_{2(g)} + 1.04\, H_2O_{(g)} \quad \Delta H^\circ = -271\ kJ/mol$$

The molar rate of creation/destruction of species j in reaction r can be written as

$$R_{rxn\,j,r} = \left(v'_{j,r} - v_{j,r}\right) K_r \prod_l \left[C_{l,r}\right]^{\eta_{l,r}} \tag{5.27}$$

where $C_{l,r}$ is the molar concentration of each reactant in reaction r (kmol/m³); $\eta_{l,r}$ is an exponent for reactant in reaction r; $v'_{j,r}, v_{j,r}$ are the stoichiometric coefficients for the jth species as product and reactant, respectively in reaction r; K_r is the kinetic rate constant for reaction r, where $K_r = A_r\, e^{(-E_r/RT)}$, (m³/kmol-s); A_r is a preexponential factor for gas-phase reaction r (m³/kmol-s); and E_r is the activation energy for gas-phase reaction r (J/kmol).

The effective rate of gas-phase combustion under the conditions prevailing in PC fired boilers may not be equal to the intrinsic kinetic reaction rate because of possible limitations imposed by mixing. When the intrinsic rate of gas-phase reactions is much higher than the rate of mixing of oxygen and combusting species, then the effective rate is controlled by the rate of mixing. Magnussen and Hjertager (1976) have proposed the eddy-dissipation model to represent the interaction between turbulent mixing and intrinsic chemical reactions. Using this model, the rate of production of species j due to reaction r, $R_{EBU\,j,r}$, is given as

$$R_{EBU\,j,r} = A\rho\frac{\varepsilon}{k}\min\left(\frac{Y_{ox}}{v_{j,r}}, Y_{fuel}\right) \tag{5.28}$$

where Y_{ox} is the mass fraction of the oxidant and Y_{fuel} of the fuel reactant. In Equation (5.28), the chemical reaction rate is governed by the large-eddy mixing time scale, k/ε, as in the eddy-breakup model of Spalding (1969). The gas-phase reaction rate is evaluated based on the minimum of the reaction rate estimated by an Arrhenius-type kinetic rate model and the eddy-dissipation model.

$$R_{j,r} - \min(R_{rxn\,j,r}, R_{EBU\,j,r}) \tag{5.29}$$

The values of the mixing time scale (k/ε) were obtained from CFD simulations, and a typical distribution of the same in a PC fired boiler is shown in Figure 5.15. The zone-specific representative values of k/ε required to estimate effective rates in each zone can be obtained either by using a point value at the center of the zone or by applying suitable averaging (line, area, and volume) over the simulated distribution obtained from the CFD model. In the illustrative example discussed here, the values of k/ε at the center of each zone were used for evaluating the effective rate using Equation (5.28).

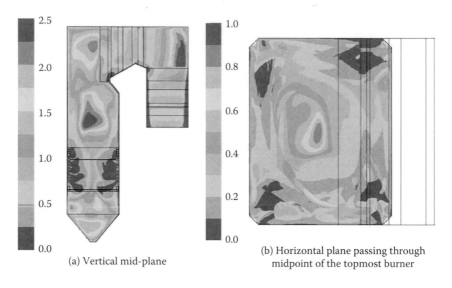

(a) Vertical mid-plane

(b) Horizontal plane passing through midpoint of the topmost burner

FIGURE 5.15 (See Color Insert.)
Typical contour of k/ε in a PC fired boiler from CFD simulation results.

5.3.4 Radiation

The most usual numerical methods for analyzing radiative heat transfer are the Monte Carlo method, heat flux method, and zone method [see Viskanta and Menguc (1987), Gosman et al. (1973), Howell (1968), Hottel and Sarofim (1967), and Hottel and Cohen (1958)]. In this study, the zone method was employed for predicting the temperature and heat flux on the water walls of the boiler. Hottel and Cohen (1958) developed this method for analyzing the radiation heat transfer in an enclosure containing gray gas with certain properties. Later, Hottel and Sarofim (1967) used this method for more complex geometries. This model has been widely used by researchers for modeling industrial radiative enclosures such as boiler furnaces (Diez et al., 2005; Batu and Selçuk, 2002). In this method, the whole space of the furnace is split into zones, and the enclosure's walls are divided into finite surface zones. The main assumption is to associate and use uniform temperatures and properties within the volume and surface zones. The heat transfer between a pair of zones depends on coefficients called heat exchange areas (see following discussion). The radiative source term is a balance between total heat exchanged by any zone with other surface and volume zones and total emission from the existing zone.

The net radiative heat source for any zone k can be written as

$$S_{rad,k} = \left(\sum_i \overline{G_k S_i} E_{s,i} + \sum_i \overline{G_k G_i} E_{g,i} \right) - 4K_k V_k E_k \qquad (5.30)$$

To evaluate the radiative exchange between the zones, it is necessary to calculate the total exchange areas $\overline{G_k S_i}$ and $\overline{G_k G_i}$. Hottel and Sarofim (1967) have discussed a detailed method to determine the total exchange area. The E_g and E_s are the black emissive power of the gas and surface, respectively. The total exchange areas are evaluated from the direct exchange areas. The direct exchange area can be calculated as

$$g_k s_j = \int\limits_{A_j} \int\limits_{V_k} \frac{K \cos\theta_j \, e^{(-kr_{kj})}}{\pi r_{kj}^2} \, dV_k \, dA_j \tag{5.31}$$

$$g_k g_j = \int\limits_{V_k} \int\limits_{V_j} \frac{K^2 e^{(-kr_{kj})}}{\pi r_{kj}^2} \, dV_k \, dV_j \tag{5.32}$$

The total exchange for one zone with all other zones is given by

$$\sum_{i \neq k} g_k g_i + \sum_j g_k s_j = 4KV_{g,k} \tag{5.33}$$

In these equations, r_{ij} is the center-to-center distance between two zones, θ is the angle between the normal vector of the surface element and the above-mentioned vector, and K is the absorption coefficient of the gas. The values of K were taken from CFD results or can be estimated using the WSGGM based on the composition, temperature, and pressure. Details about the WSGGM model were discussed in Chapter 4.

The flow rate and the temperature of air and coal at the inlet are listed in Table 4.5 in Chapter 4. The kinetic parameters for gas-phase reactions are listed in Table 4.6 in Chapter 4. The other details of a typical PC fired boiler considered here (i.e., heat exchanger surface area, tube emissivity, porosity of each heat exchanger zone, etc.) are discussed in Chapter 4.

5.3.5 Solution of Model Equations

RNM equations are solved to obtain the steady-state performance of PC fired boilers. The balance equations for the continuous and particle phases are written in the form of ordinary differential equations (ODEs) in time. For the network, the reactor domain was broken down into zones, and each zone was further divided into a series of internal reactors. The number of internal reactors for each zone will be defined by the user. For the illustrative case considered here, the equations solved for the established RNM were as follows:

- Gas phase:
  - Number of ODEs for each internal reactor = 6 (5 species + Energy)
  - Total zones = 31 (Fixed)

- • Internal reactors in each zone = N_i (User input)
- • Total number of ODEs (gas phase) = $6 \times \Sigma N_i$

- • Particle phase:
  - • Number of particle sizes = 10
  - • Number of ODEs for each size class = 4 (3 species + Energy)
  - • Number of trajectories = 4
  - • Total number of ODEs (particle phase) = 160 (= $10 \times 4 \times 4$)

Any suitable ODE solver can be used to solve these ODEs of the RNM. In this illustrative example, the modified Gear's method implemented in ODEPACK was used to solve ordinary differential equations using the LSODE (Livermore Solver for Ordinary Differential Equations) subroutine.

It should be noted, however, that the interest lies in the steady-state solution; the particle phase modeled in the Lagrangian framework will never exhibit steady state. The steady state will, however, be established in terms of the continuous phase. This particular aspect and strong interaction between the particle and the continuous phase require the development of a specific algorithm to realize effective and optimal steady-state solution of an RNM. The algorithm used for solving this illustrative example is shown in Figure 5.16.

The preprocessor reads all the relevant input data and formulates model equations of the RNM. Initial composition and temperature field are then specified for all zones. The particle-phase equations are then solved for all the considered particle streams. Please note that these equations are solved up to a specified residence time for each particle stream in all zones. Based on the simulated trajectories, the source terms for each zone of the RNM, because of the particle phase, are then calculated (for example, using Equation (5.24)). The continuous-phase governing equations are then solved using the sources calculated from the solution of the particle-phase equations. To reduce the burden on computational resources, the particle-phase sources are not updated at each time step; rather, they are updated after every time interval of τ_I. The value of τ_I can be optimized to enhance the overall effectiveness. The solution of continuous-phase equations for time interval τ_I is called one iteration of the continuous phase. After completing an iteration of the continuous phase, a check is made to examine whether or not steady state has been achieved. Various convergence criteria based on root mean square (RMS) error in the gas temperature, heat transfer to heat exchangers, and mass as well as heat balance across each zone/reactor were examined for establishing the steady state. If the steady state is not achieved, particle phase equations are solved for updating particle-phase

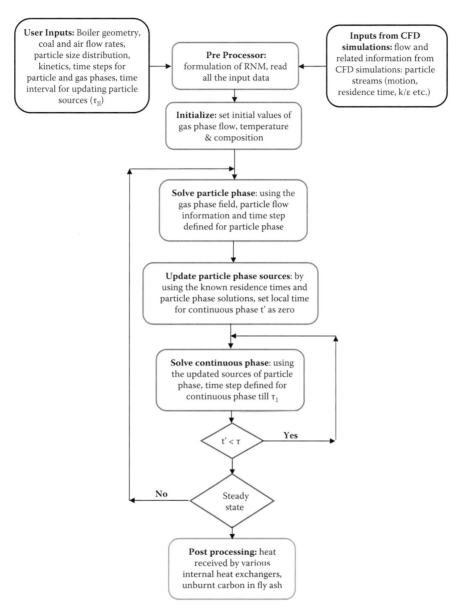

FIGURE 5.16
Solution methodology.

sources. The next iteration of continuous-phase equations is then executed using the updated particle-phase sources. In this illustrative example, we set τ_I as the mean residence time of the gas phase. The steady-state solution was achieved in typically fifteen to twenty iterations. A typical reduction in the RMS error in temperature with respect to the number of iterations is

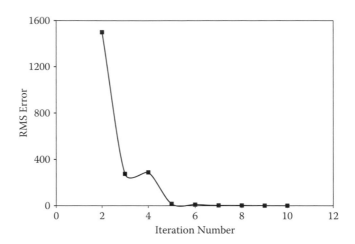

FIGURE 5.17
Convergence plot for RMS error in the temperature of zones.

shown in Figure 5.17. It can be seen that about ten iterations are adequate to obtain converged results.

After establishing the model equations and corresponding solution methods, the model was converted to an easy-to-use software tool called BOST (Boiler Simulation Tool) by Gupta (2011). An illustrative application of BOST to a typical 210-MWe PC fired boiler is discussed in Section 5.4.

5.4 Application of the RNM to a Typical 210-MWe PC Fired Boiler

The BOST was used to simulate a typical 210-MWe boiler used for illustrating application of the CFD model in Chapter 4. The underlying RNM was formulated using the CFD results. The model equations and solution method presented in this chapter were used first to simulate the base case considered for the CFD simulations. Some of the results predicted using the RNM are compared with the CFD results. The RNM was then used to simulate the influence of burner tilt to illustrate possible applications. The temperature distribution within the boiler is the key parameter of interest for simulating the performance of PC fired boilers. The simulated results of temperature within the PC fired boiler are therefore discussed here.

Unlike the CFD model, which results in a detailed prediction of the temperature field within the PC fired boiler (typically represented by contour

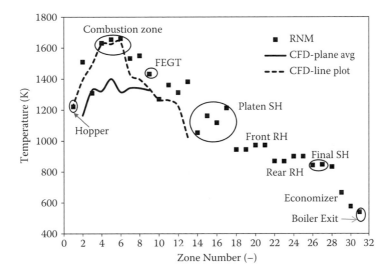

FIGURE 5.18
Gas temperature along various zones of a PC fired boiler.

plots as shown in Figure 4.23), the RNM provides the mean temperatures of different zones. The predicted temperatures of various zones in the considered PC fired boiler are shown in Figure 5.18. Some of the results obtained from the CFD field are also included in this figure. For zones in which the internal number of reactors was greater than 1, the number average of temperatures of internal reactors in that zone was estimated and plotted in Figure 5.18.

The first zone is the hopper, which shows a temperature of 1,221K. The second and third zones are the gap (2) and core (3) below the combustion zone. Zones 4 and 5 have temperatures of 1,630K and 1,654K, respectively. The combustion zone has the highest temperature, as expected. Beyond the combustion zone, the mean temperatures of subsequent zones gradually decrease until boiler exit. It can be seen from Figure 5.18 that the RNM adequately captures the trends observed from the CFD simulations with significantly less computational effort.

Part of the flow from the combustion zone travels downward from zone 4 to 2 and is further redistributed into zones 3 and 1. The predicted temperature profiles within the zones below the combustion zone are shown in Figure 5.19. The corresponding results obtained with the CFD model are also shown in this figure. It can be seen that a comparison of the results simulated using the CFD model and RNM is adequate.

The simulated temperature profiles in the combustion zone are shown in Figure 5.20(b). It can be seen that temperatures in this zone are between

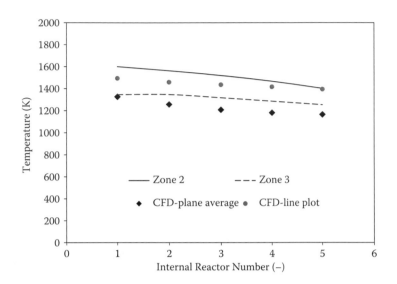

FIGURE 5.19
Gas temperature profiles in the combustion bottom zone.

1,450K and 1,750K. The temperature of the burner jet has increased from the inlet feed temperature to 1,444K in the first reactor, indicating that a significant portion of the reactions is taking place in the first reactor of zone 6. The temperature has finally increased to 1,756K. This compares well with the localized temperature across the burner jet predicted in CFD simulations (i.e., around 1,750K). The results obtained with the RNM can therefore be said to be in good agreement with those obtained with the CFD model.

The temperature profiles for zone 7 and 8 (zones above the combustion zone) are shown in Figure 5.20(a). The furnace exit gas temperature (FEGT) was measured at the outlet of the nose zone (9) as 1,432K, which is in broad agreement with the predictions of the CFD model (area weighted plane average temperature at furnace exit location).

After establishing the basic temperature distribution in and around the combustion zone, it is of interest to examine whether the RNM can capture the temperature deviation typically observed at the crossover pass of a tangentially fired PC boiler. This temperature deviation occurs mainly because of the imbalanced mass flow distribution in the crossover pass of the boiler as discussed in Chapter 4. Based on the CFD analysis, the imbalance in the mass flow rate was estimated and the same distribution was incorporated in the present RNM. The predicted temperature deviation in the crossover pass is shown in Figures 5.21(a) and (b). The temperature deviations in the crossover pass estimated using CFD and RNM models are compared with each

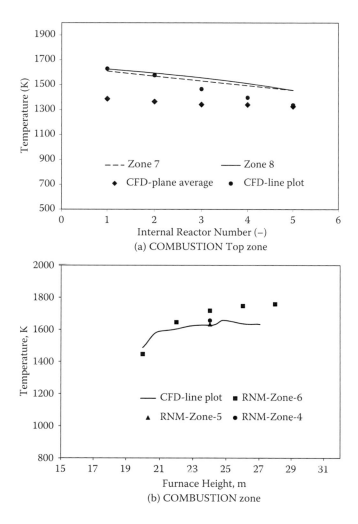

FIGURE 5.20
Gas temperature profile predicted using RNM: (a) COMBUSTION Top zone, (b) COMBUSTION zone.

other in Figure 5.21(c). It can be seen that the agreement between the values predicted using CFD and RNM is good and adequate. The results can be also improved by increasing the number of internal zones (in parallel direction to flow) of the crossover pass.

The temperature profile of the second pass zone from the LTSH to the lower economizer is shown in Figure 5.22. Each of these zones has five internal reactors, and the gas temperature along the flow direction for each internal reactor is shown in Figure 5.22. The boiler exit temperature was found to be 540K. The predicted values of heat transferred to various heat exchangers are listed in Table 5.2.

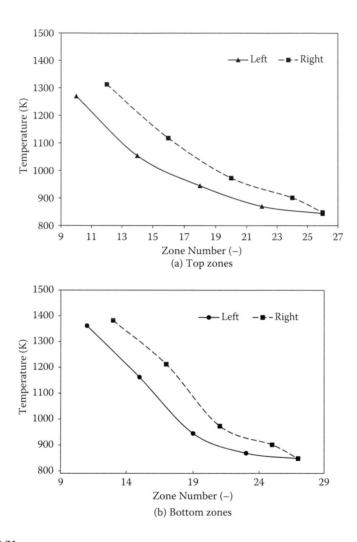

FIGURE 5.21

Typical temperature profiles and temperature deviation predicted at the crossover pass using the RNM: (a) top zones, (b) bottom zones, and (c) a comparison of the average temperature deviation predicted by CFD and the RNM. (*continued*)

A comparison shows some deviation among the values of heat transferred to individual heat exchangers predicted with an RNM and CFD model. Possible reasons for this deviation, in addition to that inherent to the model formulation, could be some differences in the surface area calculations incorporated in the CFD model and the RNM. As discussed in Chapter 4, for CFD simulations, the water walls were modeled as a flat surface, and the emissivity was adjusted to obtain the required FEGT. In the RNM, the surface area for heat exchange in each zone was estimated from the actual dimensions of

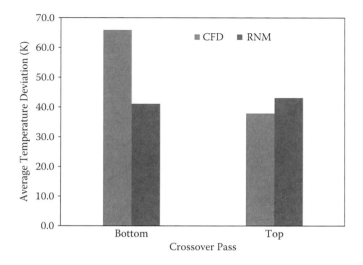

FIGURE 5.21 (*continued*)
Typical temperature profiles and temperature deviation predicted at the crossover pass using the RNM: (c) a comparison of the average temperature deviation predicted by CFD and the RNM.

the tubes of membrane panels available from engineering drawings and the boiler manual. However, the emissivity value was the same as that used in CFD simulations, and it was not tuned. In light of the different effective heat exchanger areas considered in the RNM, appropriate tuning of the effective emissivity of the tube surfaces considered in the RNM may lead to better

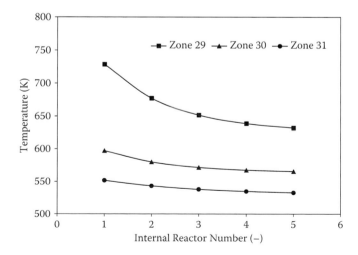

FIGURE 5.22
Temperature profile of the second pass of a boiler LTSH (zone 29), UECO (zone 30), and LECO (zone 31).

TABLE 5.2

Simulated Values of Heat Transferred to Internal Heat
Exchangers of a PC Fired Boiler

Heat Exchanger	Percentage of Heat Transferred	
	CFD	RNM (BOST)
Water wall	39	49
Platen SH	22	20
Front RH	13	8
Rear RH	6	4
Final SH	4	2
LTSH	8	11
Upper ECO	6	4
Lower ECO	3	2

Note: Total heat transferred: CFD = 540 MW; RNM (BOST) = 538 MW.

agreement. Despite some of the observed differences, the RNM can be considered successful in capturing the overall behavior of PC fired boilers.

To illustrate possible applications of the RNM, it (implemented in BOST) was then used to evaluate the influence of burner tilt on the overall performance of the PC fired boiler. The influence of tilt was tested at two conditions: +20° and −20° burner tilt. The predicted results are shown in Figure 5.23. It can be seen that the model was able to predict the movement of a hot zone as a function of tilt. For zero tilt, the burner top and boundary limits are 19 and 28 m and have a temperature of 1,654K. For +20° tilt, the hot zone has shifted in the upward direction to 22 and 33 m. This influence can be observed on the numbered average temperature of the bottom section in Figure 5.23(a), which shows that the zone 3 temperature will decrease by 126K when the burner tilt is changed from −20° to +20°. The results obtained with the RNM are compared with those obtained with the CFD model in Figure 5.23(b). It can be seen that agreement between the RNM and the CFD model is adequate for engineering purposes. The RNM can therefore be used to evaluate the influence of various key operating parameters on the overall performance of PC fired boilers. The computational time required for obtaining steady-state solution for the RNM is typically at least one order of magnitude (more likely, two orders of magnitude) lower than that required for the CFD model. The turnaround time of the RNM results may be further reduced using faster ODE solvers and better linearization of the particle source terms.

The accuracy of the RNM depends on the results of the CFD model on which the RNM is developed. Information about the flow distribution, particle residence time, tilt correlation, etc. used in the RNM is specific to a particular geometric configuration and operating conditions of a PC fired boiler (i.e., generation capacity of the boiler). The proposed approach and methodology to develop the RNM can be extended and used in a straightforward manner for other configurations (and generation capacity) of boilers provided

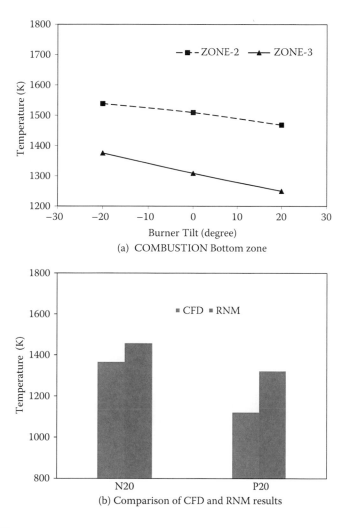

FIGURE 5.23

Influence of burner tilt on temperature profiles in the combustion bottom zones: (a) combustion bottom zone and (b) a comparison of CFD model and RNM results.

that corresponding CFD results are available. The methodology for obtaining these CFD results for any new PC fired boiler is discussed in Chapter 4, and for developing the RNM based on these results is discussed in this chapter. In some cases, even when CFD results are not available, and assuming generic development of the RNM for a particular type of boiler, RNM can be used as a first-cut simulation to explore a wide range of parameter space. Once a suitable process window is identified via these first-cut simulations, a detailed CFD model can be developed to obtain CFD results. This, in turn, can be used to carry out post-facto verification of the approximations used while developing the RNM or suitably modifying those approximations to develop the RNM. Application of models and methodologies discussed in this and the previous chapter to practice is discussed in Chapter 6.

5.5 Summary and Conclusions

Reactor network models (RNMs) of PC fired boilers require lower computational resources and offer quick turnaround compared to CFD models of PC fired boilers. In this chapter, development of the RNM based on the simulated results obtained with the CFD model was discussed. In the presented approach, the development of the RNM was based on experience, engineering judgment, and a detailed understanding of the CFD results. It requires manual inspection of the CFD results and manual formulation of zones/reactor network because at this stage it may be difficult to fully automate the formulation of the RNM. Eventually, automated formulation of the RNM from CFD simulations may be practically feasible and better than the approach proposed here. The relevant references for possible development of such automated formulations of the RNM from CFD simulations were cited.

The suggested approach for developing an RNM was illustrated by considering the case of a typical 210-MWe boiler. The CFD results for this system were discussed in Chapter 4. Using these results, RNM development was discussed here. The developed RNM was shown to predict key features of PC fired boilers with reasonable accuracy. Temperature deviation, which is one of the key factors in reducing thermal efficiency, was captured quite adequately. The RNM predictions can be further improved based on rigorous calibration with CFD results/measured point value data of gas-phase temperatures. The number of reactors needs to increase across the flow path of the gas in the crossover pass. Such an effort can provide the correlation to appropriately quantify the temperature deviation observed in PC fired boilers.

The RNM was also shown to capture the heat transferred to various internal heat exchangers reasonably well. It is possible to tune the apparent emissivity of the water wall and heat exchanging surfaces of the tube bundles and further refine the modeling of radiation included in the presented RNM to improve agreement with CFD results and plant data. The existing structure of an RNM can be improved upon by developing methodologies to automate the creation of zones from CFD simulation results, estimate the radiation exchange area coefficients using the Monte Carlo method, and further rigorous validation of plant data.

The RNM was also shown to capture the influence of burner tilt on combustion zones, and the overall performance was quite adequate. The RNM therefore has the potential to be used as a model-based on-line optimization and a process control platform. The RNM can also be used to quickly explore and evaluate large parameter spaces to identify useful "windows" of parameters offering better PC fired boiler performance. The RNM can also be used to evaluate possible changes in fuel composition (different grade coal or blends of different grade coals). The RNM, if wrapped in a user-friendly tool such as BOST, may reveal excellent benefits for computational modeling, even to the non-expert user. We hope that the discussion here will stimulate further development of such "industry-deployable" tools that will eventually facilitate the realization of performance enhancement of PC boilers in practice. Further discussion of application of CFD models and RNMs to simulate industrial PC fired boilers is included in Chapter 6.

References

Badzioch, S. and Hawksley, P.G.W. (1970). Kinetics of thermal decomposition of pulverized coal particles, *Ind. Eng. Chem. Process Design and Development*, 9, 521–530.

Batu, A. and Selçuk, N. (2002). Modeling of radiative heat transfer in the freeboard of a fluidized bed combustor using the zone method of analysis, *Turkish J. Eng. Environ. Sci.*, 26, 49.

Baum, M.M. and Street P.J. (1970), Predicting the combustion behavior of coal particles, *Combust. Sci. and Technol.*, 3(5), 231–243.

Bezzo, F., Macchietto, S., and Pantelides, C.C. (2003). General hybrid multizonal CFD approach for bioreactor modeling, *AIChE J.*, 49, 2133.

Díez, L.I., Cortés, C., and Campo, A. (2005). Modelling of pulverized coal boilers: Review and validation of on-line simulation techniques, *Appl. Therm. Eng.*, 25 (10), 1516–1533.

Field, M.A. (1969). Rate of combustion of size graded fractions of char from a low-rank coal between 1200K and 2000K, *Combustion and Flame*, 13, 237–252.

Gosman, A.D. and Lockwood, F.C. (1973). Incorporation of a flux model for radiation into a finite-difference procedure for furnace calculations, *Symp. (Int.) on Combustion*, 14, 661–671.

Gosman, A.D., Pun, W.M., Ramchal, A.K., Spalding, D.B., and Wolfshtein, M. (1969). *Heat and Mass Transfer in Recirculating Flows*, Academic Press, London.

Gupta, D.F. (2011). Modeling of Coal Fired Boiler, PhD thesis, University of Pune, India.

Gupta, D.F. and Ranade, V.V. (2008). Phenomenological Modeling of Coal Fired Boiler Based on CFD Smulations, ISCRE20, Kyoto, Japan.

Hottel, H.C. and Cohen, E.S. (1958). Radiant heat exchange in a gas-filled enclosure: Allowance for nonuniformity of gas temperature, *AIChE J.*, 4(1), 14.

Hottel, H.C. and Sarofim, A.F. (1967). *Radiative Transfer*, McGraw-Hill, New York.

Howell, J.R. (1968). *Application of Monte Carlo to Heat Transfer Problems, Advances in Heat Transfer*, Academic Press, New York.

Lowe, A., Wall, T.F., and Stewart, I.M.(1975) A zoned heat transfer model of a large tangentially fired pulverized coal boiler, *Symp. (Int.) on Combustion*, 15, 1261–1270.

Mann, R. and Mavros, P. (1982). Analysis of unsteady tracer dispersion and mixing in a stirred vessel using interconnected network of ideal flow zones. In *Proc. 4th European Conf. on Mixing*, B3, Nordwijkerhout, The Netherlands, pp. 35–48.

Magnussen, B.F. and Hjertager, B.H. (1976). On Mathematical Models of Turbulent Combustion with Special Emphasis on Soot Formation and Combustion. In *16th Symp. (Int.) on Combustion,* The Combustion Institute.

Morsi, S.A. and Alexander, A.J. (1972). An investigation of particle trajectories in two-phase flow systems, *J. Fluid Mech.,* 55(2), 193–208.

Mujumdar, K. and Ranade, V.V. (2009). Modeling of Germany, Rotary Cement Kilns, VDG Verlag.

ODEPACK LIVERMORE Solver, http://computation.llnl.gov/casc/odepack/ode-pack_home.html.

Osawe, M., Felix, P., Syamlal, M., Lapshin, I., Clectus, K.J., and Zitney, S.E. (2002). An integrated process simulation and CFD environment using the CAPE-OPEN interface specifications, *Proc. of AIChE Annual Meeting,* Indianapolis, IN.

Ranade, V.V. (2002). *Computational Flow Modeling for Chemical Reactor Engineering,* Academic Press, London.

Ranade, V.V., Chaudhari, R.V., and Gunjal, P.R. (2011). *Trickle Bed Reactors,* Elsevier, Amsterdam, The Netherlands.

Ranz, W.E. and Marshall, W.R. (1952). Evaporation from drops. I., *Chem. Eng. Progr.,* 48, 141–146.

Ray, H.W. and Wells, G.J. (2005). Methodology for modeling detailed imperfect mixing effects in complex reactors, *AIChE J.,* 51(5), 1508.

Smith, I.W. (1971). Kinetics of combustion of size-graded pulverized fuels in the temperature range 1200–2270°K, *Combustion and Flame,* 17(3), 1971.

Wang, Y.-D. and Mann, R. (1992). Partial segregation in stirred batch reactors, *Chem. Eng. Res. Des.,* 70A, 282–290.

6

Application to Practice

The objective of this chapter is to provide guidelines to practicing engineers on how to develop and use computational models of pulverized coal (PC) fired boilers. Previous chapters discussed computational flow dynamics (CFD) as well as phenomenological models in detail. The scope here is restricted to offering additional comments for facilitating applications of these models to practice. An attempt is also made to include common pitfalls while developing and applying computational models to PC fired boilers.

6.1 Performance Enhancement Using Computational Models

Typical performance enhancement using computational models consists of the following steps:

1. Clearly identify specific performance enhancement objectives.
2. Relate the performance objectives to underlying processes and determine "performance-controlling" processes (sometimes the objective of modeling is to determine performance-controlling processes: "learning" models as discussed in Chapter 2).
3. Set up a "wish list" for computational models.
4. Formulate computational models (model equations, boundary conditions, etc.).
5. Map the computational model to a "solver" in order to enable simulations.
6. Verify and validate using limiting solutions/available industrial data.
7. Apply computational models to gain an understanding of existing scenario.
8. Use that understanding to evolve creative solutions for realizing performance enhancement.

9. Use the models to evaluate alternative solutions and identify the most promising one.

10. Implement the solution and verify performance enhancement. If not satisfactory, go to Step 2.

Identifying "performance-controlling" processes (Step 2) and then setting up the "wish list" (Step 3) for the computational model are the most important steps. The success of the application of mathematical (or otherwise) modeling in any performance enhancement project depends on setting up such "wish lists"; they act as a map or a guide for the selection and application of relevant models. The results obtained by these various models are then used to evolve suitable performance enhancement solutions.

A typical wish list for enhancing the performance of PC fired boilers could be

- Increase efficiency (less coal/MWe).
- Get more throughput per unit volume (compact boiler).
- Have a more environmentally friendly operation (lower NOx and SOx).
- Reduce downtime of boiler (by reducing cleaning time and frequency).

The next step is to translate this performance enhancement wish list into a quantitative form and establish a relationship between items on this wish list and the degrees of freedom available (boiler hardware and operating protocols). It may often turn out that some of the items on the wish list require contradictory options of hardware and operation. In such a case, a careful analysis of the different items on the wish list must be made to assign priorities. Operability, stability, and environmental constraints often take precedence over efficiency and throughput when such conflicting requirements arise. With these general comments, specific guidelines for using computational flow modeling (CFM) and phenomenological models in practice are discussed in the following two sections.

6.2 Application of CFD Models to PC Fired Boilers

The development and application of CFD models to PC fired boilers (or, for that matter, to any equipment) requires several steps. These may be broadly grouped into four categories:

1. Formulation of a CFD model (pre-processing)
2. Solution of model equations (solver)

3. Examination and interpretation of simulated results (post-processing)

4. Use of simulated results for performance enhancement (application)

The key aspects of these categories are briefly discussed in the following.

6.2.1 Formulation of CFD Models

6.2.1.1 Selection of a Solution Domain

The first step in formulating a CFD model of a PC boiler is to decide the solution domain to consider and then model the geometry of the selected solution domain. The selection of the solution domain itself is not easy. The solution domain should be selected in such a way that it captures the key region of interest and allows specification of the influence of the outer environment (not considered in the model) on the boundaries of the selected solution domain.

In most PC fired boilers, part of the air fed to the boiler is actually fed to the mills from which air entrains PC particles and carries them to the boiler. Therefore, it may be tempting to consider the solution domain right from the mills (see "Solution Domain A" shown schematically in Figure 6.1). However, more often than not, the solution domain of a typical boiler simulation excludes mills and inlet ducts, and usually starts from the burner faces opening into the boiler furnace. If there is specific interest in quantifying the behavior of the inlet ducts, it is recommended to formulate a separate CFD

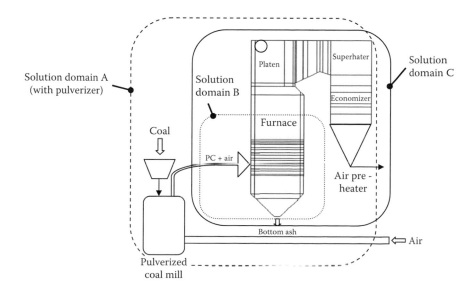

FIGURE 6.1
Solution domains for simulating pulverized coal boiler.

model for carrying out flow simulations from the mill to the burner face opening into the boiler furnace (see, for example, Bhasker, 2002).

It is also important to appropriately select the outer boundary of the solution domain. If the interest is restricted to the furnace, it may be in principle possible to reduce the size of the solution domain by excluding most of the superheaters and other heat exchangers embedded in the PC fired boiler (see domain B shown in Figure 6.1). The other extreme is to consider the solution domain right up to the inlet of downstream equipment, such as filters and/or electrostatic precipitators. However, the solution domain C shown in Figure 6.1 is usually recommended. This domain includes embedded superheaters and other heat exchangers because their presence significantly influences flow and heat transfer within the boiler. It ends after the economizers, as shown in Figure 6.1. This choice of solution domain not only covers all the regions of interest, but also allows the formulation of appropriate boundary conditions at the boundaries of the solution domain in a relatively straightforward way.

If the specific objectives of developing a CFD model are different from the overall simulation of a PC fired boiler, it may be essential to select different solution domains. These different cases are not discussed here for the sake of brevity. However, an overall approach can be used to select appropriate solution domains with reference to the objectives.

6.2.1.2 Geometry Modeling and Mesh Generation

After identifying the solution domain, the next step is to computationally model the geometry of the considered domain. Anyone familiar with the fabrication drawings of a PC fired boiler will know the geometrical complexity of the boiler. It contains thousands of tubes, supports, bolts, louvers, welded joints, manholes, platforms, etc. It is almost impossible to capture all these details in a single computational geometry. Fortunately, it is not essential to consider each and every component/geometrical complexity while formulating a CFD model of a PC fired boiler.

It is possible and usually necessary to simplify the geometrical complexity. The usual rule of thumb is to ignore the geometrical intricacies if they are not influencing the flow and heat transfer processes of interest. However, for modeling PC fired boilers, there are several areas where geometrical intricacies will influence the flow and heat transfer, and must still be simplified in order to keep the model computationally tractable. It is then essential to account for these simplifications in the model formulation. It is therefore recommended *not* to start directly from the fabrication drawings of the boiler. Rather, it is advisable to make conscious decisions about possible simplifications and build the boiler geometry from scratch as per those simplifications. Some usual and useful simplifications are briefly outlined here.

Water walls of the boiler: The furnace walls of a PC fired boiler are lined with water tubes in which steam is generated. The dimension of such tubes (60 to 90 mm) is usually quite small compared to the overall width of the furnace (10 to 25 m). Resolution of each individual tube therefore will enormously increase the demands on computational resources. It is therefore advisable not to resolve individual tubes and model water walls as a flat surface. It should, however, be mentioned that the effective heat transfer area of the modeled domain will be much smaller than the real heat transfer area. This must be accounted for while formulating the flow and heat transfer model. One of the ways to adjust this is with the help of the apparent emissivity of the water wall (this value is adjusted to account for changes in real and modeled heat transfer areas). This is discussed in Chapter 4.

Embedded heat exchangers in the boiler: There are several heat exchangers in the typical solution domain of the boiler. The diameters tubes used in such heat exchangers are, quite small compared to the size of the overall solution domain. Therefore, it is recommended to model these internal heat exchangers as porous volumes rather than resolving individual heat exchanger tubes. This approach was discussed in Chapter 4. The characteristics of the hypothetical porous volumes representing internal heat exchangers must be formulated in such a way that the formulation accurately captures the flow and heat transfer characteristics of actual heat exchangers (see Chapter 4 for more details).

In addition to these and similar approximations, it is recommended to account for mesh generation requirements while modeling the geometry of PC fired boilers. It is recommended to split the geometry into several different volumes to provide appropriate control of mesh generation, as well as to facilitate the formulation of region-specific models to account for local approximations. Usually, PC fired boiler geometry is modeled by defining distinct regions such as the bottom hopper, combustion chamber, nose, crossover pass, superheater, economizer, etc. Each of these regions may have to be split into different slices. For example, the combustion chamber may contain several subregions that represent different burners and other air inlets. Aligning the dimensions of such subregions with the dimensions of the burners and other smaller openings of the tangentially fired boiler helps during the process of mesh generation.

After simplifying and modeling the geometry of the considered solution domain, the next task is to generate the appropriate mesh. Critical analysis of the prevailing characteristic velocity, length, and space scales within the considered solution domain is a prerequisite for initiating mesh generation. Quantification of these characteristic scales will provide generic guidelines for selecting the appropriate number of computational cells. A typical rule of thumb is that the size of the computational cells should be smaller by at least one order of magnitude than that of the smallest relevant length scale of interest. Characteristic length scales of course may vary within the solution domain, and therefore an estimation of this variation is useful for exploring possible ways to reduce the mesh count.

After broadly estimating demands on the resolution of different regions, generally it is possible to begin the meshing process from any section of the geometry. For PC fired boilers, the combustion zone is more complex as it has many burners, a complex flow structure, and the majority of heat release takes place in this region. Therefore, it is usually recommended to start meshing from the burner regions. Almost all state-of-the-art CFD tools allow the use of unstructured grids. However, it is still recommended to use hexahedral meshes (rather than tetrahedral meshes) wherever possible. Recent versions of CFD tools also accept the use of polyhedral meshes. Mesh generation in and of itself is a vast subject area, and several excellent books and reviews on this are available (for example, please see Frey and George, 2008; Thompson et al., 1997; and Liseikin, 2010). Readers are referred to these resources and references cited therein for more details. Here we include only a few generic comments.

Despite the challenges in keeping the mesh count to a reasonable number, users should take care to avoid extremes of aspect ratios, skewness, and volume ratios of adjacent computational cells. Typical numbers of computational cells used in recent studies are listed in Table 6.1. It is recommended to spend significant time and effort on grid generation to save time and resources at later stages in the simulations/project. It is also necessary to formulate grid sequencing and refinement strategies to understand the influence of grid spacing/distribution on simulated results. More often than not, it will be difficult to obtain truly grid-independent solutions for PC fired boilers. Systematic grid sequencing studies may help to derive maximum benefit from the simulated results, despite the nonavailability of the truly grid-independent solution. The grid sequencing also helps in convergence and overall resources required for simulations. Most state-of-the-art mesh generation tools allow parametric grid generation. These facilities can be used to reduce the time spent on mesh generation. Quite often, some refining operations on generated mesh are needed to facilitate convergence. Some of these include (1) grid smoothing (includes operations that adjust node locations while maintaining element connectivity) and (2) grid clean-up (operations that change element connectivity). It is worthwhile to spend the effort on understanding tools for grid refinement and tools for assessing the quality of the generated grid to generate desired meshes.

6.2.1.3 Formulation of Boundary Conditions

After meshing the geometry, the next step is to define boundary conditions on all the boundaries of the solution domain. This is not as simple as it sounds.

Air inlets/burner inlets: Air entering the burner is usually divided into three groups; primary air (PA), secondary air (SA), and over fire air (OFA). It is important to specify the velocity, composition, and temperature of the incoming streams. In a typical tangentially fired boiler, jets entering the

TABLE 6.1

Number of Computational Cells Used for Simulations of PC Fired Boilers

Ref.	Boiler Capacity (MWe)	Tangential (T)/ Wall Fired (W)/ W-Shaped	Furnace Dimensions (W×D×H)	Number of Computational Cells
Fang et al. (2012)	200	T-fired	11.92×10.8×44.6	809,868
Gupta (2011)	210	T-fired	10×13×52	1,465,013
Diez et al. (2008)	600	T-fired	n. a.	517,000
Asotani et al. (2008)	40	T-fired	5.5×5.5×17.5	278,640
Yin et al. (2002)	609	T-fired	n. a.	454,776
Al-Abbas et al. (2012)	550	T-fired wall-mounted burners	17.82×17.82×98.84	559,006
Tian et al. (2009)	375	W-fired	15.9×15.9×77.5	950,000
Belosevic et al. (2008)	350	T-fired wall-mounted burners	15 1×15.1×43	549,250
Vuthaluru and Vuthaluru (2006)	500	W-fired	19.5×18×58	192,950
Coda and Tognotti (2000)	320	W-fired	10×10×30 (furnace height)	64,750
Xu et al. (2001)	300	W-fired	15×11.43×47	140,700
Fan et al. (2001)	300	W shaped	21×15.6×43	292,400
Xu et al. (2000)	350	W-fired	17.13×10.67×48.7	108,630
Fan et al. (1999)	600	W-fired	17×18×57.075	185,185

Note: n.a.: not available.

boiler are not necessarily horizontal. It is therefore essential to obtain the following information to specify these inlet boundary conditions:

- Distribution of air flow rates as PA, SA, and OFA (note that flow rates will be calculated by considering the appropriate pressure and temperature at the inlet).
- Distribution of each type of air into the burner slot mounted on the corner. Due to the ducting layout, it is possible to have nonuniform distribution of the air into each burner located at the corner of the furnace. It is also possible to formulate a separate CFD model of ducts in order to estimate possible maldistribution. In the absence of this information, uniform distribution of air may be assumed at the first instance. At a later stage, sensitivity studies can be performed to estimate the impact of possible maldistribution.
- Angle at which these jets are injected into the boiler from the corner of the furnace.

After obtaining this information, three velocity components of air at each inlet face are calculated and specified to the CFD solver. It is important to note that appropriate temperature and pressure values estimated at the inlet faces must be used for converting mass flow rates to volumetric flow rates/ velocities. In addition to these, many times swirling flow is used in burners to create effective turbulent mixing near the burner tip, which enhances combustion of the volatiles and char. The swirl components must be specified accurately while formulating inlet boundary conditions. If necessary, a separate CFD model that resolves all the details of burner vanes may have to be developed to estimate the extent of swirl accurately.

In addition to characteristics of the air, it is also essential to specify appropriate boundary conditions for the coal particles. Key characteristics required to be specified include

- Mass flow rate of coal (and volume fraction of coal particles in some cases)
- Particle size/size distribution of coal particles
- Inlet velocity of coal particles
- Temperature of coal particles
- Composition of coal particles

It should be noted that coal particles generally get heated because of grinding. It is important to understand key characteristics of pulverizing mills for correctly specifying boundary conditions for the coal particles. Usually, coal particles are assumed to enter the boiler with the same velocity as that of the air. The particle size distribution of coal particles entering the boiler is often represented by a Rosin–Rammler distribution. It is often necessary to capture the particle size distribution (PSD) accurately. If the PSD is not adequately represented by a Rosin–Rammler distribution, alternative distributions such as log-normal may have to be used to represent coal particles (see Crowe et al., 2012, for more details on different distribution functions that can be used for representing PSD).

Water wall boundary conditions: As mentioned earlier, water walls are usually modeled as a flat surface, and an appropriately corrected value of the emissivity must be specified at these walls. As water flowing through the water walls is almost at saturated condition, the inlet and outlet temperatures of fluids passing through the water wall remain constant. Hence, the input condition to the water wall can be specified as a constant-temperature wall. This assumption simplifies the overall formulation of boundary conditions. Another key boundary condition at water walls is regarding coal particles. The usual approach is to specify a restitution coefficient at the walls, which will ensure that impinging particles will rebound from the walls. On some occasions, depending on the temperature of the wall as well as the composition of the particle, it may be possible that impinging particles will stick to the walls. Appropriate boundary conditions using specially developed, user-defined functions may have to be implemented in such cases.

Outlet boundary conditions: Usually, a constant-pressure outlet boundary condition is used. It should be remembered that, in reality, the boiler operates under a slight vacuum. This vacuum may result in ingress of ambient air from the surroundings if there are any leaks. It is important to examine and estimate this possibility of air ingress to ensure that simulated boundary conditions represent reality. For solid particles, a standard escape boundary condition may be specified at the bottom hopper as well as at the outlet. Data about distribution of the ash in the bottom and top of the boiler are usually available from the plant or from the boiler manual. The specified boundary conditions and implemented model should be able to correctly represent this known distribution of ash (bottom ash and fly ash). If the simulated results indicate lower particle capture in the bottom than that observed in the plant, then boundary conditions at the bottom hopper may have to be adjusted (adjust the area through which particles are exiting the solution domain from the bottom hopper). This correction is very important as it can directly affect the thermal efficiency prediction of the model.

In addition to specifying boundary conditions at the real boundaries, it is also essential to specify the appropriate characteristics of porous volumes (to estimate pressure drop and heat transfer coefficients) representing various internal heat exchangers. These parameters can be estimated by carrying out separate CFD simulations, as discussed in Chapter 4. These estimated characteristics can then be provided to each heat exchanger block using a user-defined function (UDF).

6.2.1.4 Specifying Physical Properties

It is important to specify appropriate auxiliary equations to estimate—as accurately as possible—the physical properties and their dependence on temperature as well as composition. Most commercial CFD codes solve enthalpy equations and calculate temperatures from enthalpy values using specified auxiliary equations for specific heats. Errors in the estimation of specific heats may result in inaccurate estimations of the temperature. This will lead to significant errors in radiative heat transfer as well as in rates of some of the chemical reactions. Every effort should therefore be made to develop and specify accurate auxiliary equations to estimate the effective heat capacity, density, conductivity, etc. Because the flow is usually turbulent, estimation of the molecular viscosity is not as critical as estimation of the heat capacity. Usually, the ideal gas law provides adequate estimation of gas-phase density.

In addition to heat capacity, another important property is the absorption coefficient and emissivity of the gas. For a correct representation of the temperature field, the local absorption coefficient should be estimated based on the concentrations of water and CO_2 as well as the temperature of the gas. The weighted sum of gray gas model (WSGGM) is recommended for estimating the absorption coefficient, which provides an optimum option between the simplified gray gas model and the banded gray gas model. Our numerical experiments have shown that there is no significant influence of the value of the scattering coefficient on the overall heat transfer in the boiler. The value of the scattering coefficient may therefore be set to zero. The wall emissivity value is generally set in the range of 0.6 to 0.8 for PC boiler simulations.

6.2.1.5 Turbulence and Two-Phase Models

The overall flow in the boiler is turbulent. Appropriate turbulence models therefore must be specified. A brief summary of typical turbulence models used for the CFD simulation of PC boilers was presented in Chapter 4. Usually, the standard k-ε model is used to simulate turbulent flow in PC boilers at least at the first instance. If there is significant swirl at the burners, an RNG or realizable k-ε model may be recommended.

As discussed in Chapter 4, the Eulerian–Lagrangian approach is usually used for simulating PC boilers. Appropriate inlet boundary conditions, including realistic particle-size distribution (represented by, say, the Rosin–Rammler distribution), need to be specified. Details of choosing different submodels within the Eulerian–Lagrangian framework are discussed by Ranade (2002) and may be referred to for selecting appropriate settings.

6.2.1.6 Chemical Reactions

Key gas-phase and gas-solid reactions taking place in PC fired boilers were discussed in Chapters 3 and 4. Most commercial CFD codes provide various options/models for specifying homogeneous and heterogeneous reaction kinetics. Reaction kinetics are often related to the rank/type of the coal. Generally, such data are difficult to obtain from the open literature. Key reaction parameters (at least for devolatilization) can be obtained from thermogravimetric analysis (TGA) or drop-tube furnace experiments, as discussed in Chapter 3. For combustion reactions, different approaches have been used to formulate kinetic models such as global kinetics (Baum and Street, 1970), intrinsic kinetics (Smith, 1971), and coupled extinction phenomena-based (Suuberg, 1991 and Hurt et al., 1998) approaches. Generally, simple Arrhenius-type kinetic models are used. The parameters of such models can be obtained from drop-tube experiments (Gupta, 2011, Williams et al., 2002; Arrenilas et al., 2002; Backreedy et al., 1999).

Gas-phase reactions are modeled using either a finite-rate chemistry model or a mixture fraction/probability density function-based approach. The finite-rate chemistry-based approach solves transport equations for the reactant and product concentrations where the chemical reaction mechanism

is specifically defined. Gas-phase reactions in a boiler occur at high temperatures and therefore the effective reaction rate is often influenced by turbulent
mixing. A simple eddy-dissipation approach, presented by Magnussen and
Hjertager (1976), is often used in CFD simulations. In this approach, rates
estimated based on intrinsic kinetics and turbulent mixing are compared,
and the slower of the two is selected for simulations. This approach thus
naturally implements intrinsic kinetics at lower temperatures, and uses turbulent mixing beyond a certain critical temperature. Measurements of the
fireball temperature (and the size of fireball) in the PC boiler, if available, can
be used for indirect validation of specified kinetic models.

6.2.1.7 Heat Transfer

The heat generated due to all the reactions discussed above should be transferred to the heat exchangers provided to the furnace of a boiler in order to
control the gas and particle temperatures. In the combustion zone, radiation is the predominant mode of heat transfer. Radiation may account for
almost 70% of the heat transfer in a PC fired boiler. Accurate modeling of the
radiation is therefore crucial to capture flow, temperature, and composition
distribution in the boiler. It is important to include the influence of participating media (e.g., ash and coal particles) on radiative heat transfer. For large
systems such as PC fired boilers, P-1 and discrete ordinate (DO) models are
usually used.

The selection of an appropriate thermal radiation model for a PC fired
boiler depends on the model's capability of efficient handling participating
media and the localized heat source, and being computationally efficient for
optically thick media. The P-1 model works well for systems with a large
optical thickness (>3 m). Optical thickness is a product of the absorption coefficient and characteristic length scale. The DO model works across the range
of optical thicknesses. However, it is computationally more expensive than
the P-1 model. A comparative study performed on a 160-MWe boiler indicates that the DO model with forty-eight flux directions (accuracy increases
with an increase in flux directions) will take approximately twice the time
that a P-1 model will take for convergence (Filkoski, 2010). The simulated
results, however, do not show a significant difference in the predictions of
these two models. The discussion and some results presented in Chapter
4 also indicate that there is no significant variation in the simulated results
with either the DO or P-1 model. The P-1 model is therefore generally recommended for simulating PC fired boilers.

Convective heat transfer is estimated based on the boundary conditions
provided to the water wall in the simulation. Because the water walls are
modeled as constant-temperature heat sinks, the local heat transfer coefficients are estimated during the solution. For the suspended heat exchangers, the heat transfer coefficient is estimated based on the correlation for

convective heat transfer. Such correlations are estimated based on the periodic heat transfer studies performed on unit cells. For estimating convective heat transfer to these porous volumes, the effective temperature difference is calculated using the gas temperature of the current cell and tube metal wall temperature which is calculated as an average of the inlet and outlet temperature of the steam water of the particular heat exchanger can be considered.

Similarly, the radiative heat transfer to porous volumes is also modeled using the gas temperature of the current cell and tube metal wall temperature. In reality, the heat exchanger surface can exchange heat with multiple surfaces and gas volumes at distant locations based on the view factor, and can interact with reflected and scattered rays as well. However, because these heat exchangers are modeled as porous volumes and not as tube bundles, there are limitations to calculating such radiative heat transfer rigorously. Nevertheless, it is possible to write user-defined functions that can account for such intrinsic details if someone is interested in estimating the wall temperatures of embedded heat exchangers. This will, however, significantly enhance convergence difficulties. The deposition on the heat exchanger surface increases the surface temperature of the deposits and hence reduces the effective heat transfer to the heat exchangers. This also increases the flue gas temperature exiting the boiler. Implementation of a particle deposition model can predict the surface temperature due to particle deposition; and based on these inputs, CFD simulation can be re-run with revised boundary conditions for the heat exchangers.

6.2.2 Solution of Model Equations

After formulating model equations and boundary conditions, the next important task is to select appropriate numerical methods and algorithms to solve the model equations and specify various numerical parameters required to apply selected methods and algorithms. In addition to these, it is also important to carry out simulations sequentially:

- From coarse grids to finer grids
- From simplest single-phase isothermal flow to multiphase flow with combustion and heat transfer

The overall methodology is shown in Figure 6.2. It is recommended to start with isothermal simulations on a coarse grid. The influence of the number of computational cells and turbulence models on simulated flow fields can be examined from these simulations. In many cases, cold air velocity data in the boiler under consideration are available. Such data can be used to identify the appropriate mesh count and turbulence model. Other measurable parameters such as pressure drop can also be obtained from such simulations, which can be used to evaluate developed CFD models.

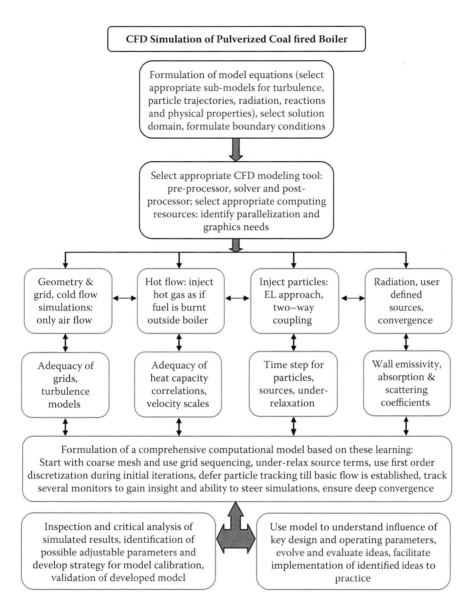

FIGURE 6.2
Overall solution methodology.

Once the flow field is established, the particles under isothermal conditions can be injected into the furnace. This helps establish particle trajectories in the furnace/boiler. It is important to ensure that an adequate number of particle trajectories are simulated for a sufficient time in order to ensure that most of the particles injected into the boiler exit the solution domain.

After establishing particle trajectories, reactions can be activated. It may be useful to patch approximate temperatures and compositions to initiate reactive simulations.

The convergence of combined simulations of flow, heat transfer, and reactions is tricky. Several tricks, as well as a systematic methodology, should be developed to achieve convergence quickly. Some of the key tips are outlined in the following:

- Initiate simulations with coarse grids and first-order discretization. Switch to higher-order discretization schemes after obtaining reasonable convergence.

- Specify appropriate under-relaxation parameters: it is often useful to have lower under-relaxation parameters for momentum equations so as to reduce errors in the continuity equations. It should be noted that errors in continuity equations may result in spurious sources or sinks in the species conservation equations. Therefore, even at the expense of a slower rate of convergence, it is recommended that one use lower than usual under-relaxation parameters for the momentum equations (that is, 0.1 to 0.2).

- The time step used in simulating particle trajectories should be chosen in such a way that particles generally cross, at most, one cell within one time step.

- Source terms, because of the particle-level processes, should be under-relaxed as well. More importantly, these sources must be linearized appropriately while implementing in the CFD model to ensure smooth convergence.

- Source terms representing embedded heat exchangers may also have to be linearized and under-relaxed appropriately.

To obtain a better understanding of and insight into radiative heat transfer, sometimes numerical experiments can be performed by assuming that all the energy content of the injected coal is already liberated at the inlet and by setting a corresponding temperature of the gas at the inlet. Such numerical experiments may allow for a quick check of key parameters of the radiative heat transfer and may also facilitate selection of appropriate models. Submodels for estimating absorption (constant or WSGGM), scattering coefficients, and wall emissivity can also be tested with such studies.

For simulating the combustion of injected coal particles, it may be useful to initially patch a high-temperature zone in the furnace region (e.g., 1,200°C to 1,300°C). This provides the necessary activation temperature to initiate the combustion reaction and may facilitate convergence as well as a steady-state solution. Initiation of solid combustion generates very large source terms for mass as well as for energy equations of the gas phase. It is therefore advisable to use smaller under-relaxation factors (URFs) for solid particle source terms and radiation equations at the beginning of particle combustion simulations. As the solution progresses, each of these URFs can be increased to higher values (eventually close to 1). Another important parameter that determines stability and overall convergence rate is the frequency at which the particle source term is updated while solving the gas-phase equations. Usually, computation of solid-phase sources requires significantly large computing resources (as they are based on simulations of several particle trajectories). It is therefore recommended that particle source terms are updated after a few iterations of the gas-phase equations (on the order of ten iterations at least, and typically may be thirty to fifty iterations).

Determining the convergence is also crucial. In general, it is recommended to use nondimensionalized residuals of the equations being solved as one of the criteria. Most of the CFD codes, by default, report normalized residuals that are often not a very good indicator of errors. In addition to residuals, it is also essential to monitor certain macroscopic quantities to judge the extent and approach to convergence. Errors in macroscopic mass and energy balance, mass weighted furnace exit gas temperature (FEGT), and heat transferred to water walls and to various heat exchangers embedded in the PC fired boiler are natural choices for monitors. After establishing that the iterative solution of model equations is converged, various post-processing tools can be used to inspect the simulated results. Some of these aspects are outlined in Section 6.2.3.

6.2.3 Examination and Interpretation of Simulated Results

Each simulation of a PC fired boiler generates tens of millions of numbers. It is important to develop effective post-processing strategies and methodologies to make the best use of these simulated results. The key purpose of CFD simulations is to gain insight, rather than just to get some numbers. It is therefore important to start developing post-processing strategies by reiterating the objectives behind carrying out simulations and key questions on which insight is expected. The state-of-the-art CFD tools offer several excellent post-processing tools consisting of the following:

- *Contour plots:* These are mainly used to examine and display the distribution of various scalar quantities such as temperature, velocity magnitude, species concentrations, etc. on any plane within the solution domain. These are quite useful in quickly exhibiting key features of the solution.
- *Vector plots:* A velocity field is conventionally inspected via vector plots where the velocity field is represented by arrows. The length

of the arrows is proportional to the velocity magnitude, and arrows point in the direction of the velocity.

- *Iso-surfaces:* Most post-processing tools provide several additional features beyond contour and vector plots. An iso-surface is a three-dimensional surface within the domain representing loci of constant values of relevant variables.

Particle trajectories: Particle trajectories can also be inspected to understand the motion of particles within the domain.

Most of these post-processing features allow various combinations and overlays to display more information via a single picture. For example, particle trajectories (variation in particle coordinates with respect to time) may be colored by, for example, the temperature of the particle or the remaining carbon content of the particle. Such combinations often reveal significant information that otherwise is unavailable. Particle trajectories can also be colored using the gas-phase variables, such as oxygen concentrations in the gas phase. This type of analysis can clearly bring out regions in which oxygen might be deficient (than the stoichiometric requirement). Similarly, iso-surfaces can also be colored by a variety of gas-phase and particle-phase variables.

The general methodology is to take slices of the three-dimensional (3D) model at various heights from bottom of the boiler to the top of the boiler and plot contours of the desired variables on each of these planes. Such sectional plots can also be observed at the center of the furnace depth. The burner section temperature plots can provide information about the temperature profile of the jets entering from each burner. Significant efforts are usually spent on understanding the intricacies of flow, heat transfer, and reactions in PC fired boilers, especially in the fireball region (furnace). Observations on the release of combustion products while coal particles travel in the jet are also important. This information is useful in locating temperature hotspots in the combustion zone. Sources of impurity generation within the solution domain can be examined using either contour plots or iso-surfaces of impurity mass fractions or impurity generation rates. The FEGT is an important parameter; it is a function of furnace design and is generally kept between 100°C and 150°C below the ash deformation temperature. Hence, this temperature can be monitored using the mass average of a plane at the nose exit of the furnace.

Most post-processes associated with CFD tools also allow users to probe the simulated results, either by reviewing alphanumeric data or using traditional two-dimensional (x-y) plots. Classical x-y plots are often useful in evaluating the possible influence of numerical aspects of simulations. Model calibration and validation are often carried out using such x-y plots. A large variety of alphanumeric data can be inspected. Apart from the main variables used and stored by the CFD solver, it is also possible and quite useful to define and examine "user-defined" variables while post-processing. Macroscopic balances of mass, species, and energy are very useful for understanding the overall behavior of simulated PC fired boilers. The use of some of these post-processing tools was demonstrated in Chapter 4 while discussing a sample of results simulated using the CFD model described in that chapter.

The particle residence time distribution and average particle residence time in the furnace can be obtained from particle trajectory data. The residence time in the different zones of PC fired boilers, along with the variation in particle composition, can also be inspected. These results may provide better insight into particle-level processes and their interactions with gas-phase flow. The information on the distribution of solids in the bottom and top of the boiler is useful in estimating the relative ratios of fly ash and bottom ash.

The post-processing of simulated results is carried out first to ensure that simulations are not unduly influenced by numerical issues and do capture reality. Further analysis of simulated results focuses on developing a better understanding of the various processes occurring in PC fired boilers. The gained understanding is expected to facilitate an evolution of creative ideas on performance enhancement. The models can then be used to further evaluate these ideas and identify the most promising options for practical implementation. Some aspects of such applications of simulated results for performance enhancement are outlined in Section 6.2.4.

6.2.4 Application of Simulated Results for Performance Enhancement

After obtaining converged results and inspecting them via a variety of post-processing tools as discussed in the previous section, the next step is to verify and validate the obtained results. A classical analyst's paradox states that "everyone believes an experiment except the experimentalist and no one believes an analysis except the analyst." The onus of establishing the adequacy and usefulness of computational modeling is therefore on the modeler (or analyst). Verification is ensuring the "right" solution of model equations; it does not depend on the equations being solved. It is mainly based on truncation errors: the numerical solution should represent the exact solution as mesh size tends to zero. Validation is ensuring that the "right equations" are solved; it addresses questions such as

- Do the equations being solved truly represent the physical system under consideration?
- Are there any assumptions that are physically incorrect?
- Is the nature of flow (compressible, transient, etc.) represented correctly?
- Are the size of the domain and boundary conditions appropriate?

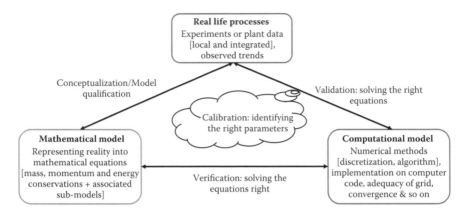

FIGURE 6.3
Verification and validation.

This is illustrated in Figure 6.3. The computational flow modeling is never one pass unless prior fine-tuning is made. Even for a well-converged case, based on the information available in post-processing, the analyst often goes back to the grid or solver for

- Higher-order schemes/refined mesh: to quantify discretization errors
- Post-facto verification of adequacy of mesh, wall treatment, and boundary conditions
- Evaluating some of the underlying assumptions made while formulating model equations

In the true sense, a numerical analysis can only be validated with experiments. The objective of performing validation is to estimate errors in simulated results, taking into account all the associated factors. The definitions of the errors and uncertainties associated with numerical simulations have been the subject of significant research. Interested readers can refer to Freitas (2002), Oberkampfa and Trucano (2002), and Roache (1997) for more details.

More often than not, systematic error and uncertainty analysis may not be practical for large and complex applications such as PC fired boilers because of several constraints on available resources (time, experimental data, computing resources, cost, etc.). Appropriate strategies therefore must be developed to make meaningful inferences from the simulations and to use the simulated results for realizing performance enhancement in practice. Usually, the first step in evaluating simulations of a PC fired boiler is to compare simulated values of macro-characteristics with the available data/ boiler design manuals. Typically, key variables include FEGT, gross heat transferred to each heat exchanging tube bundle, temperature at a few local points in the crossover pass and second pass, and oxygen concentration after the economizer obtained from simulations are evaluated. Any furnace-level data on the temperature distribution within the furnace and combustion section (can be measured with charge-coupled device [CCD] infrared camera), if available, are very useful in evaluating computational models.

If some critical differences (differences larger than possible error bars associated with data and numerical analysis) are observed between simulated results and plant data, the first step is to "verify" that the simulations are correctly solving the equations. Although full rigorous methods for verification may not be practically feasible for PC fired boiler simulations, strategies such as grid sequencing and extrapolation to zero grid size are often used. After verifying that the solution is correct, attention then focuses on whether the equations being solved are correct. The underlying physics used for formulating models to simulate PC fired boilers are very complex, and therefore ideally one should individually validate the various submodels used in the overall computational model. However, resources for carrying out this exercise are very rarely available. Knowledge and insight based on past experiences and engineering judgment are therefore used for developing the models. With the limited degree of freedom available to develop models, the alternative approach, albeit with limited validity, is to "calibrate" the model.

Model calibration focuses on identifying the correct parameters for the chosen model equations. Computational models of complex systems such as PC fired boilers use several different model parameters (see, for example, the models described in Chapter 4). The values of these parameters often influence the simulated results. The appropriate choice of such model parameters is therefore very crucial. The art of modeling lies in identifying a smaller set of model parameters that can be adjusted to reduce the observed gap between the simulated results and reality. One of the ways to select these parameters is to carry out systematic sensitivity analyses with respect to various parameters and identify the most relevant parameters based on these results. Usually, an attempt is made to identify independent parameters to adjust for differences observed in the key variables of interest. Possible choices of adjustable parameters for reducing the gap between simulated results and plant data in the case of PC fired boilers are listed in Table 6.2.

After establishing reasonable agreement between simulated results and available plant data/experience, the computational model can then be used to gain better insight. For example, aspects of fireball formation, its shape and size, peak temperature, etc. can be investigated in detail. The temperature gradients experienced by injected particles in the furnace region (temperature–time history) can also be examined. The temperature deviation observed in the crossover pass can be quantified. This deviation occurs mainly because of the mass flow imbalance at the entrance to the crossover pass due to swirling flow. The insights gained through critical analyses of simulated results can be used to evolve ideas for performance enhancement. For example, for eliminating or reducing temperature deviation (in

TABLE 6.2

Possible Choices for Model Calibration

No.	Observed Differences in Simulated Results and Data	Parameters that May Be Adjusted to Reduce This Gap
1	Overall pressure drop, pressure drop across internal heat exchangers	Turbulence models, porous volume representation of internal heat exchangers (porosity, viscous and inertial coefficients)
2	Heat transferred to internal heat exchangers	Absorption coefficient, scattering coefficient, parameters of WSGG model, surface emissivity or effective surface area (for porous volume approach), surface temperatures of internal heat exchangers (given as inputs from water side calculations), heat loss via bottom ash (please see point 5 for this), Nusselt number parameters for convective heat transfer rate
3	Heat transferred to water walls	Emissivity of water walls and appropriate effective surface area, absorption coefficient of gases, heat loss via bottom ash, heat loss to the atmosphere
4	Local hot spots	Heat capacity of gases
5	Distribution of ash (fly ash to bottom ash)	Particle restitution coefficient, particle boundary conditions (rebound, escape, trap, and so on)
6	Unburned carbon in ash	Kinetic parameters, kinetic models (possibility to account for extinction phenomena), possibility of air ingress, extent of excess air
7	Shape, size, and temperature of fireball	Burner geometry, turbulence model, kinetics of char combustion, homogeneous reactions in gas phase, extent of excess air
8	Pollutant concentration	Fireball temperature (see points 3 and 6), NO_x: Sensitivity of parameters like N_2 distribution in the volatiles & Char and N_2 into HCN & NH_3
9	Temperature deviation at crossover path	Burner geometry, turbulence models + see points 4 and 5
10	Flue gas compositions	Air ingress, extent of excess air, kinetic parameters, pollutant formation
11	Overall thermal efficiency	See points 2 to 7

other words, mass imbalance) in the crossover pass, ideas such as modifying the geometry of the boiler nose (location, size, and shape) or superheater (changing the length or tube pitch) may be evolved and evaluated. Various ideas for reducing undesired pollutants by making certain changes in the furnace region, such as staging of air, tilting of burners, or changing the type of burner, extent of excess air and its distribution, etc., may evolve and be evaluated using the calibrated models. Some of these ideas may appear obvious in hindsight. However, it has been proven again and again that detailed information available through computerized simulations provides the most effective way of gaining insight, evolving ideas, and evaluating these ideas to shortlist the most promising ones for implementation in practice.

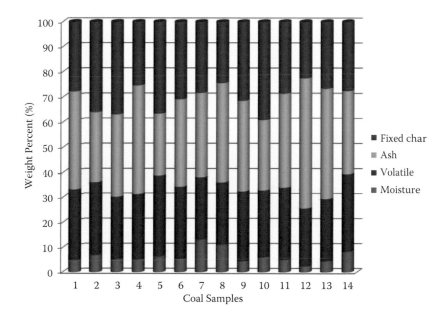

COLOR FIGURE 3.2
Typical characteristics of Indian D (Sample no. 2-5, 7-10, 14) & E (Sample no. 1, 6, 11–13) type coal (data from Mishra, 2009).

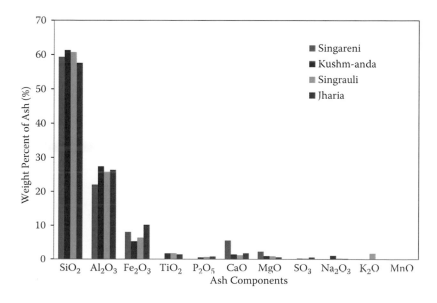

COLOR FIGURE 3.3
Typical ash composition of Indian coal (data from Chandra, 2009).

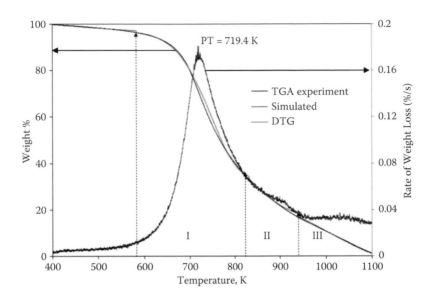

COLOR FIGURE 3.6
Typical results for devolatilization of high-ash coal (TGA: weight loss, DTG: rate of weight loss).

Moisture : ash : VM : FC :: 12 : 41 : 23 : 24
C : H : N : S : O :: 37 : 2.26 : 0.85 : 0.33 : 6.53

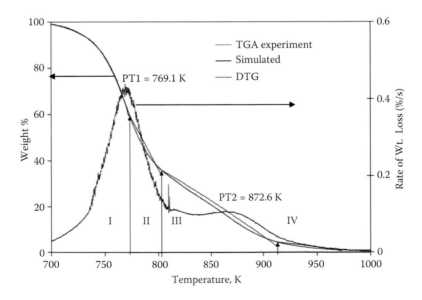

COLOR FIGURE 3.9
Results of char oxidation of a typical high-ash coal (TGA: weight loss; DTG: rate of weight loss).

Moisture : ash : VM : FC :: 12 : 41 : 23 : 24
C : H : N : S : O :: 37 : 2.26 : 0.85 : 0.33 : 6.53

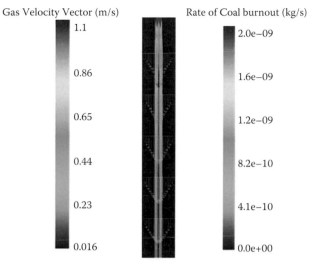

(a) 2D axi-symmetric model (1723 K)

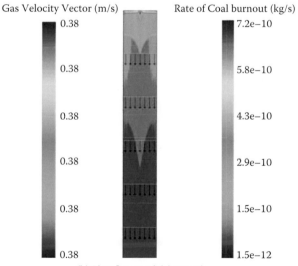

(b) Plug flow model (1723 K)

COLOR FIGURE 3.19
Distribution of char burnout rates (superimposed with velocity magnitude vectors): (a) 2D axi-symmetric model (1,723K) and (b) plug flow model (1,723K).

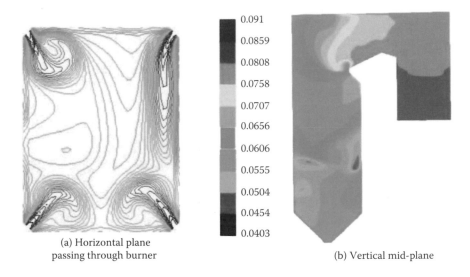

	0.091
	0.0859
	0.0808
	0.0758
	0.0707
	0.0656
	0.0606
	0.0555
	0.0504
	0.0454
	0.0403

(a) Horizontal plane
passing through burner

(b) Vertical mid-plane

COLOR FIGURE 4.4
Simulated distribution of absorption coefficient (1/m) using WSGGM: (a) horizontal plane passing through burner and (b) a vertical mid-plane.

COLOR FIGURE 4.8
Fouling observed in PC fired boilers: deposit build-up on superheater after 1 week of co-firing (coal + straw). Reprinted from *Progress in Energy and Combustion Science*, 31 (5–6) Zbogar, A., Flemming, J.F., Jensen, P.A. and Glaborg, P., Heat Transfer in Ash Deposits: A Modelling Tool Box 261–269 (2005) with permission from Elsevier.

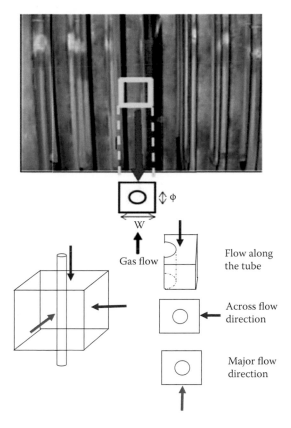

COLOR FIGURE 4.12
Representation of actual tube bundle of internal heat exchanger as a porous volume.

COLOR FIGURE 4.14
Simulated temperature distribution across a tube of a typical platen superheater.

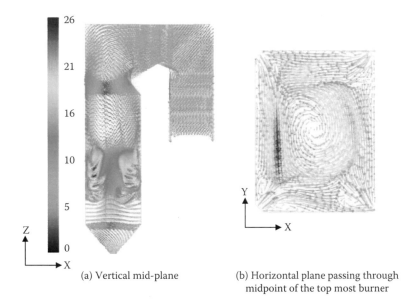

(a) Vertical mid-plane

(b) Horizontal plane passing through midpoint of the top most burner

COLOR FIGURE 4.20
Simulated gas flow field (velocity shown in m/s): (a) vertical mid-plane and (b) horizontal plane passing through the midpoint of the topmost burner.

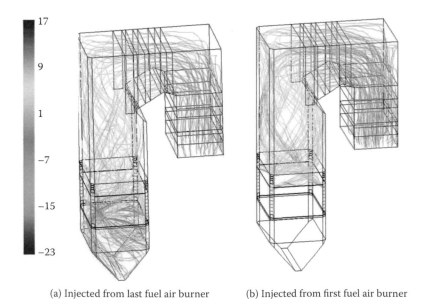

(a) Injected from last fuel air burner

(b) Injected from first fuel air burner

COLOR FIGURE 4.22
Simulated trajectories of coal particles colored by z velocity (m/s) of the particle.

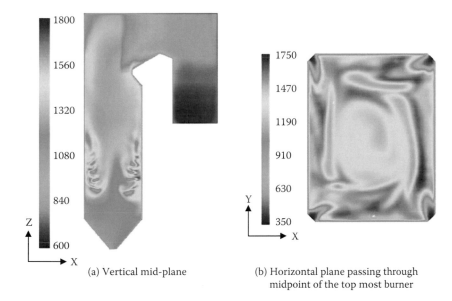

(a) Vertical mid-plane

(b) Horizontal plane passing through
midpoint of the top most burner

COLOR FIGURE 4.23
Typical temperature (K) distribution within a PC fired boiler.

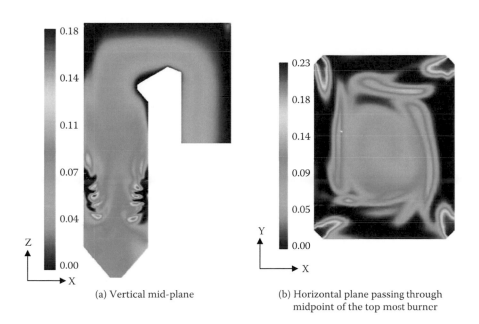

(a) Vertical mid-plane

(b) Horizontal plane passing through
midpoint of the top most burner

COLOR FIGURE 4.24
Typical distribution of O_2 mass fraction within a PC fired boiler: (a) vertical mid-plane and (b)
horizontal plane passing through the midpoint of the topmost burner.

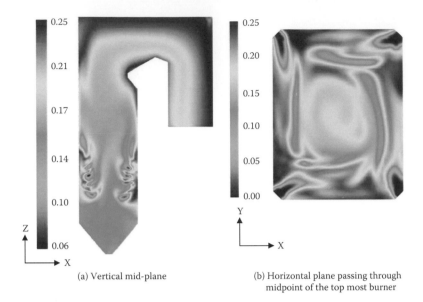

(a) Vertical mid-plane

(b) Horizontal plane passing through midpoint of the top most burner

COLOR FIGURE 4.25
Typical distribution of CO_2 mass fraction within a PC boiler.

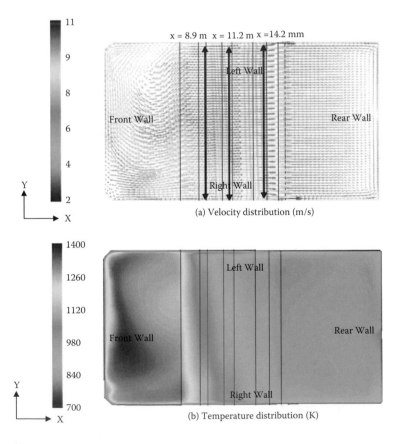

x = 8.9 m x = 11.2 m x = 14.2 mm

Left Wall

Front Wall

Rear Wall

Right Wall

(a) Velocity distribution (m/s)

Left Wall

Front Wall

Rear Wall

Right Wall

(b) Temperature distribution (K)

COLOR FIGURE 4.26
Simulated results at crossover pass. ($z = 47$ m).

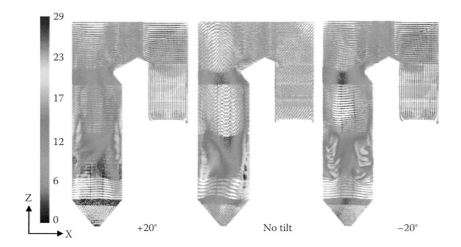

COLOR FIGURE 4.34
Influence of burner tilt on simulated velocity field (m/s) at vertical mid-plane.

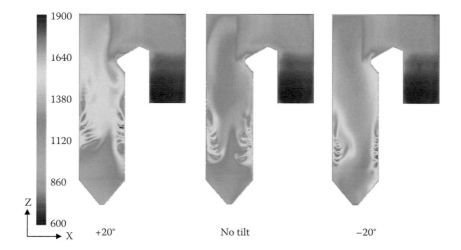

COLOR FIGURE 4.35
Influence of burner tilt on simulated temperature distribution (K) at vertical mid-plane.

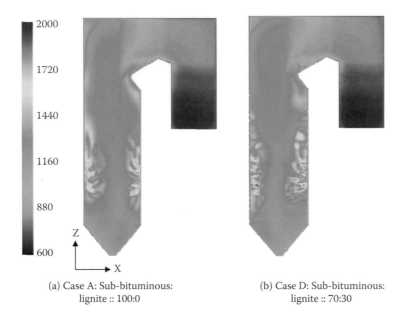

(a) Case A: Sub-bituminous:
lignite :: 100:0

(b) Case D: Sub-bituminous:
lignite :: 70:30

COLOR FIGURE 4.44
Influence of coal blends on simulated temperature (K) distributions at vertical mid-plane: (a) Case A: Sub-bituminous: lignite :: 100:0, and (b) Case D: Sub-bituminous: lignite :: 70:30.

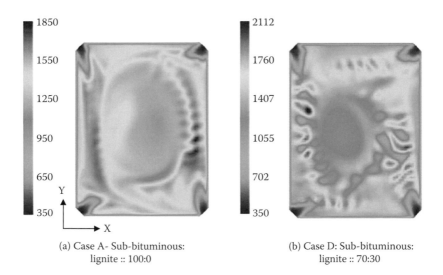

(a) Case A- Sub-bituminous:
lignite :: 100:0

(b) Case D: Sub-bituminous:
lignite :: 70:30

COLOR FIGURE 4.45
Influence of coal blends on simulated temperature (K) distributions at horizontal plane passing through the midpoint of the bottommost burner: (a) Case A: Sub-bituminous: lignite :: 100:0, and (b) Case D: Sub-bituminous: lignite :: 70:30.

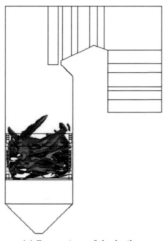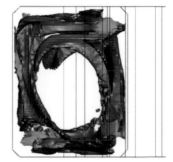

(a) Front view of the boiler   (b) Top view of the furance

COLOR FIGURE 5.3
Iso-surface of temperature in furnace zone (1,550K): (a) front view of the boiler and (b) top view of the furnace.

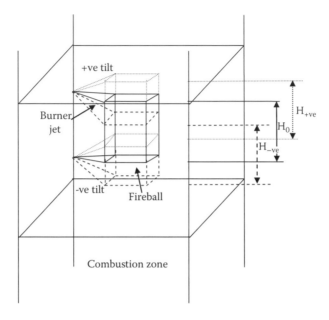

(a) Schematic of the FIREBALL zone at different burner tilts H_0, H_{-ve} and H_{+ve}: FIREBALL position at zero tilt, downward tilt and upward tilt of burner respectively.

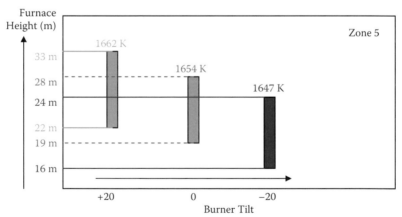

(b) Sample of influence of burner tilt on temperature and location of FIREBALL zone

COLOR FIGURE 5.5
Effect of burner tilt on fireball zone.

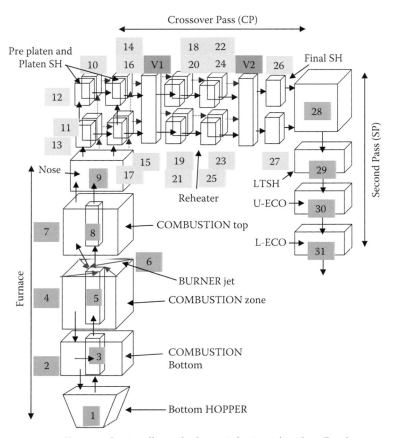

Zone number in yellow color boxes indicates right side wall and
similarly blue color indicates left side wall of the boiler,
Green arrow indicates the burner jets.

COLOR FIGURE 5.6
Reactor network model for tangentially fired PC boiler.

1: HOPPER;
2: COMBUSTION BOTTOM (CB) GAP; 3: COMBUSTION BOTTOM (CB) CORE;
4: FIREBALL GAP; 5: FIREBALL; 6: BURERJET;
7: COMBUSTION TOP (CT) GAP; 8: COMBUSTION TOP (CT) CORE;
9: NOSE; 10 TO 13: PRE PLATEN; 14 TO 17: PLATEN;
V1: VIRTUAL VOLUME; 18 TO 21: FRONT RH; 22 TO 25: REAR RH;
V2: VIRTUAL VOLUME; 26 TO 27: FINAL SH;
28: PASS2TOP; 29; LTSH; 30: U-ECO; 31:L-ECO

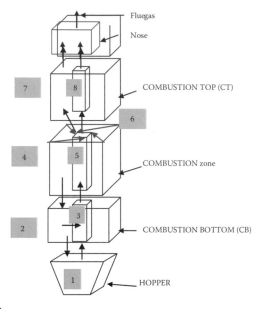

COLOR FIGURE 5.7
Schematic of a reactor network for the bottom section.

1: HOPPER;
2: COMBUSTION BOTTOM (CB) GAP; 3: COMBUSTION BOTTOM (CB) CORE;
4: FIREBALL GAP; 5: FIREBALL; 6: BURERJET;
7: COMBUSTION TOP (CT) GAP; 3: COMBUSTION TOP (CT) CORE

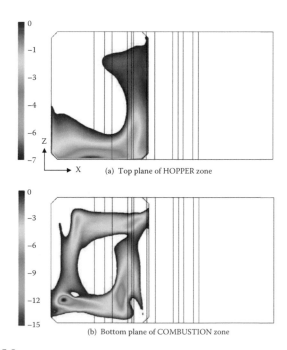

COLOR FIGURE 5.8
Region showing downward flow at two horizontal planes below the combustion zone (contours of z velocity, m/s).

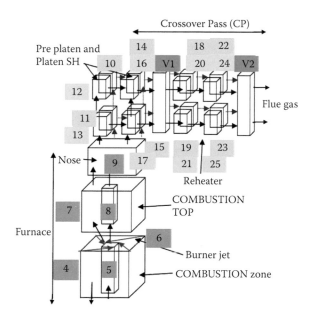

COLOR FIGURE 5.10
Reactor network for fireball-CT-nose section.
4: GAP;
5: FIREBALL;
6: BURERJET;
7: COMBUSTION TOP (CT) GAP;
8: COMBUSTION TOP (CT) CORE;
9: NOSE;
10 TO 13: PRE PLATEN;
14 TO 17: PLATEN

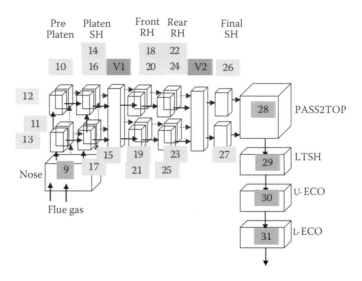

COLOR FIGURE 5.12
Reactor network for platen superheater to economizer section.

9: NOSE; 10 TO 13: PRE PLATEN; 14 TO 17: PLATEN;
V1: VIRTUAL VOLUME; 18 TO 21: FRONT RH; 22 TO 25: REAR RH;
V2: VIRTUAL VOLUME; 26 TO 27: FINAL SH;
28: PASS2TOP; 29: LTSH;30:U-ECO;31:L-ECO

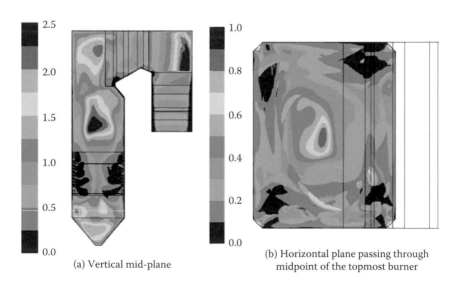

(a) Vertical mid-plane

(b) Horizontal plane passing through midpoint of the topmost burner

COLOR FIGURE 5.15
Typical contour of k/ε in a PC fired boiler from CFD simulation results.

6.3 Application of Reactor Network Models to PC Fired Boilers

The CFD-based models require large computing resources for simulating PC fired boilers. Various aspects of development and application of such models for PC fired boilers were discussed in the previous section. The CFD models provide an excellent platform for gaining better insight and stimulating ideas for performance enhancement. However, a CFD model is not an optimal platform for exploring parameter space over a wide range. Second, CFD models are also not very convenient to use for on-line optimization of PC fired boilers. It is essential to develop a lower-order model that requires much less computing resources and time and therefore is more convenient to use for on-line optimization as well as to explore a large parameter space. Development of such lower-order models (called reactor network models (RNMs)) is discussed in Chapter 5.

RNMs may help boiler designers estimate the basic furnace dimensions quickly and perform sensitivity studies on various aspects such as configuration, heat exchanger area, burner tilt, fuel properties, fuel blends, etc. RNMs allow the exploration of a wider parameter space in an effective manner. Such preliminary simulations can provide first-cut designs for a boiler that can further be explored and studied later in detail using the CFD model. This approach is better than simply relying on the rules-of-thumb or general guidelines for the sizing of the boiler. In recent years, "model-based" on-line process control has been increasingly practiced. Considering the ever-increasing emphasis on maintaining and improving PC fired boiler efficiencies, it is essential to develop computational models that can be integrated with on-line process control applications. RNMs, with their lower computational requirements, can be conveniently used with on-line process control platforms. It is important to use a systematic methodology to develop and use RNMs to realize such objectives in practice.

RNMs do not solve momentum equations. Usually, the PC fired boiler is represented by a network of reactors that exchange heat and mass with each other. The information concerning the flow (of gas as well as particles) within the PC fired boiler is either assumed based on actual measurements and experience or on simulated results obtained from the CFD model. The development and application of RNMs to PC fired boilers (or, for that matter, to any equipment) requires several steps. These may be broadly grouped into four categories:

1. Formulation of reactor network (number, size, location of reactors, and network of connections among them)
2. Formulation of model equations for the established reactor network (mass and energy balance equations for gas and particle phases)

3. Solution of model equations

4. Use of simulated results either to explore a new parameter space or for on-line control and optimization of a PC fired boiler

The key aspects of these categories are briefly discussed here.

The formulation of a reactor network is key for developing a useful RNM. In a typical CFD model, the geometry of the PC fired boiler is modeled with millions of computational cells (as discussed in Chapter 4). To reduce the demand on computational resources, RNMs typically represent the geometry of PC fired boilers with about 100 computational cells (i.e., with a network of about 100 reactors). The number, size, and location of the reactors are the basic factors to consider while developing an RNM. It should be noted that each of the reactors considered in an RNM is treated like a completely mixed zone. It is therefore essential to ensure that the reactors do not encompass regions where there is significant nonuniformity (of temperature or compositions). Either actual measurements or simulated results of the CFD model can be used to identify such pseudo-homogeneous regions within PC fired boiler geometry to facilitate formation of the RNM. The methodology of formulating an RNM based on CFD models is discussed in Chapter 5. It is usually adequate to formulate a network of thirty to eighty reactors to represent a PC fired boiler.

The next important step is to establish connections among the formulated reactors (to create a network of reactors). This is usually done either using gross mass balance based on certain assumptions or using simulated results of a CFD model. There are various methodologies for coarse-graining CFD simulations to formulate reactor network models (see discussion in Chapter 5 and Gupta (2011), Diez et al. (2005), and Bezzo et al. (2003)). Appropriate care should be taken while establishing mass flow rates along the reactor networks. It is recommended to carry out a degree-of-freedom analysis carefully to identify which variables will be considered free variables.

After establishing the reactor network, it is relatively simple and straightforward to develop basic mass and energy balance equations for the established reactor network. The particle phase is usually modeled in the Eulerian–Lagrangian framework. See Chapter 5 for details of the formulation of equations. These governing equations are usually ordinary differential equations (ODEs), and many standard methods as well as solvers are available to solve a set of ODEs and algebraic equations. Considering the strong coupling between the solid phase and the gas phase, explicit methods are not recommended. Appropriate tolerances must be specified and care must be exercised to ensure that simulated results are not sensitive to any numerical issues (viz. tolerances, time step for gas and particle phase, etc.).

The size of the simulated results of RNMs is orders of magnitude smaller than that of CFD models. The post-processing of RNM results is therefore usually carried out with conventional x-y plots. Very often, RNM-based simulations are carried out in an automated fashion to cover a wide range of parameters, and the processing of results is also usually automated using the standard tools such as MATLAB® or MS Excel®. Occasionally, contour plots (as discussed in the previous section) can also be used to examine results of the RNMs.

In view of the fact that an RNM requires significantly less computational resources to run, it is tempting to use it for quantifying the influence of various parameters over a wide range. It should, however, be kept in mind that meaningful results from an RNM can only be obtained as long as the underlying assumptions used while formulating the reactor network are valid over the considered range of parameters. The RNM can also be directly coupled with a model-based process control platform for on-line control and optimization. As on-line process control usually optimizes PC fired boiler operation by manipulating adjustable variables over a narrow range, RNMs work very well. In case there is any doubt about whether or not the underlying assumptions of an RNM are valid, it is advisable to use CFD-based models to assess the validity of the underlying assumptions or reconfigure the reactor network to suit the conditions of interest. It is also possible to couple an RNM with a CFD model and use the hybrid model for variety of applications.

An RNM or CFD model and combinations of the two offer a significant potential to develop better insight and to realize performance enhancement in practice. It is important to be aware of some of the common pitfalls in the application of computational modeling to practice. In the Section 6.4, we attempt to identify potential pitfalls in computational modeling and provide suggestions on how to avoid them.

6.4 Common Pitfalls in Computational Modeling

Various steps in developing and applying computational models were briefly discussed in Section 6.1 (as Steps 1 through 10). The starting point of the computational modeling cycle is typically a need for performance enhancement or a real-life problem that must be resolved (Step 1 in Section 6.1). The next step in the development of computational models essentially involves identifying key aspects of the target system after removing all irrelevant aspects of reality. An appropriate complexity level is selected and the corresponding model equations are formulated. Based on the model and desired objectives, the model is implemented on a computer and numerical simulations are carried out. The results of the simulation experiments are analyzed with respect to the set objectives. The results and conclusions are communicated to the stakeholders for implementation in practice. Usually, this process takes place iteratively with subsequent refinements to earlier model versions. Our experience of using computational modeling for a variety of industrial applications indicates that the entire exercise of computational modeling appears to be prone to a variety of common pitfalls in practice. In this section we would like to share some of those pitfalls that we have experienced. Recently, Barth et al. (2012) nicely summarized some of the pitfalls in modeling and simulation based on their experiences from armed forces and business. Although the subject we are discussing here is quite different from those discussed by Barth et al. (2012), we have borrowed their structure for discussing pitfalls in computational modeling of PC boilers.

Developers and users of computational modeling of PC fired boilers might encounter common pitfalls in the following stages:

- Defining objectives of the computational model
- Formulating the computational model
- Implementing the computational model
- Interpreting the simulated results
- Accepting the recommendations drawn from the simulated results

It is tempting to define broad and grand objectives for the computational model while developing the project proposal. However, it is important to clearly conceive and formulate the project with specific objectives. If the objectives are not clearly specified, addressing general objectives typically requires enlarging the part of the reality to be modeled. This might make the model structure complex and may therefore have to carry baggage in excess of that required for addressing specific objectives. It is therefore recommended to stay focused and stay away from generating a "mother-of-all" computational model or simulation. For example, it is much better to state the objective of the computational model of a PC fired boiler as "understand the causes of unburned carbon in fly ash and develop ways to minimize it" rather than "understand flow, heat transfer, and reactions in a PC fired boiler and develop a comprehensive platform for simulating the PC fired boiler." The latter objective may be set as a goal of an academic research group where a large number of Ph.D. students and post-doctoral fellows might work several years to take steps toward this goal. It is, however, not a suitable objective for realizing performance enhancement in practice.

The next common pitfall concerns selecting the appropriate level of complexity of the model to be developed. Model development is an iterative process, and there may be two approaches to arrive at the appropriate level of complexity; start with the simpler model and add complexity if the evidence suggests that the model is not adequate for capturing key features demonstrated in reality. Alternatively, start with the more complex model and remove some complexity if evidence suggests that specific features in the model are not relevant to the objective at hand. Selecting an undue complexity level leads to many other issues related to model structure, quantity and quality of input data, and required resources. As emphasized previously, a skilled modeler must understand the difference between "simple" and "simpler" models. The model must represent reality sufficiently well for the simulation to yield applicable results. However, including all the physics known up to the time of model development bears the risk of drowning in details and losing sight of the "big picture." Sometimes, this complexity even causes a simulation project to fail. The balancing act between simplification and exact representation is therefore very crucial. Simplifications involve uncertainty about modeling decisions, and people who would like to avoid such decisions are prone to this pitfall.

The main goal of a computational model is to be as realistic as possible without jeopardizing "tractability." The computational model should be judged as "useful" or "not so useful," rather than "right" or "wrong." It is therefore important to stay focused on the core cause-and-effect relationships when creating a model. It is always useful to look for possibilities to simplify the model structure in order to avoid putting its tractability and analysis at risk. The flow, heat transfer, and reactions occurring in PC fired boilers offer plenty of opportunities to trap users into this pitfall. For example, the gas flow in PC fired boilers is turbulent. Several turbulence models of different complexities are available (see Ranade, 2002). It is, however, prudent to select two-equation turbulence models for many practical applications related to PC fired boilers rather than selecting more sophisticated Reynolds stress models or large-eddy simulation models at least at the first instance.

Implementation of a computational model that involves appropriate selection of computer and software platforms to implement numerical methods for solving model equations offers the next set of common pitfalls. The cost and time pressures often propel users toward these pitfalls. The time and cost required to learn a new modeling platform or programming language can be quite high. Instead of selecting computer and software platforms that suit the problem at hand, often the existing computer and software platforms are used to implement the computational model.

After completing and testing an implemented computational model, users are prone to losing their critical attitude while analyzing the simulated results. This may lead to undiscovered model errors or reduced efforts in validation. It also should be remembered that a model is a simplified representation of reality. It is only expected to provide valid results within the context of underlying assumptions. Going beyond that context faces the danger of inappropriate or even false conclusions. For example, if the objective of the computational model of a PC fired boiler is to quantify heat transferred to various internal heat exchangers, it is usually adequate to represent the char combustion with rather simple models. Once such a computational model is developed, it will be inappropriate to use that model to understand and quantify possible unburned carbon in fly ash. That objective will require different and more sophisticated char combustion models than those that are adequate for the heat transfer applications.

To avoid this kind of pitfall, simulated results should be critically analyzed to evaluate their plausibility and to find alternative explanations for the results. It is also preferable to involve a third party (other than the user carrying out the simulations) to execute this critical analysis.

Even after carefully executing the computational modeling project and providing specific recommendations, there is a possibility that the real stakeholders will not accept these recommendations. The more complex the simulation model, the more skeptical they tend to be about the results. We have observed a typical dilemma: if the simulated results agree with the expectations, the results are termed "trivial." If the results are unexpected, then the computational model is termed "questionable." The only way to reduce the impact of this kind of pitfall is to carry out systematic verification and validation exercises, as well as avoid the impression of a "black box"-type computational model.

The common pitfalls that can be encountered while applying the computational models of PC fired boilers to practice are listed in Table 6.3. Some comments to avoid these pitfalls are also included in this table. We hope that the suggestions included in this table, and an awareness of common pitfalls and methodology/suggestions discussed in this chapter, will be useful to the readers.

6.5 Summary

An attempt was made in this chapter to provide guidelines to practicing engineers on how to develop and use computational models of PC fired boilers. Computational models, if used judiciously, often provide most effective ways for gaining insight, evolving ideas, and evaluating these ideas to shortlist the most promising ones for implementation in practice. Identifying "performance-controlling" processes and then setting up the "wish list" for the computational model is one of the most important steps. Details of the CFD model and the RNM were discussed in Chapters 4 and 5, respectively. Additional comments on the application of these models for realizing performance enhancement in practice were included in this chapter. Some common pitfalls were highlighted as well. Hopefully, these comments will be useful to practitioners desiring to use computational models in practice and will further stimulate the development as well as the application of computational models for enhancing the performance of PC fired boilers.

TABLE 6.3

Common Pitfalls in Applying Computational Models to Practice

Situation	Possible Pitfall	Comments
Defining objectives of the model	Vague definition of objectives and inadequate clarity on how model results will be used in practice	Specific objectives of computational modeling should be spelled out. Clear visualization of intended use of simulated results help to translate performance enhancement objectives into modeling objectives.
Modeling geometry of industrial PC fired boiler	Temptation to include all geometrical details	All geometrical details are not necessary and are often detrimental to overall success (may have to spend undue resources on grid generation, computing, etc.).
Selecting solution domain	To reduce computational resources, domain size is curtailed	It is essential to select domain in such a way that appropriate boundary conditions can be formulated. Compromises on this often lead to not-so-useful simulated results.
Selecting submodels: turbulence, radiation, reactions, etc.	Temptation to select the most complex model	It should be noted that a more complex model does not necessarily mean a more accurate model. More complex models also demand significantly more input parameters. Appropriate models must be selected based on available data and objectives under consideration.
Simulations with commercial CFD solvers	Temptation to include all relevant physics simultaneously at the beginning of simulations	Refer to Chap. 4 for systematic guidelines on this.
	Implementation of large computational model on a massively parallel computer	It is essential to understand the interactions among computer architecture, software implementation of parallelization algorithm, and extent of coupling and nonlinearity in model equations for making appropriate choices/selections.

(Continued)

TABLE 6.3 (*Continued*)

Common Pitfalls in Applying Computational Models to Practice

Situation	Possible Pitfall	Comments
Detecting convergence	Relying on normalized residuals	Need to ensure deep convergence based on variety of monitors and non-normalized residuals.
Interpretation of results/ assessing accuracy of simulation	Temptation to look for expected trends/results; unrealistic expectations about accuracy and validity of simulated results	Post-processing of large data sets often offers opportunities to non-objective interpretation. Care must be exercised to avoid this. Efforts should be directed toward gaining better insight. Involve other stakeholders and independent experts to review interpretation of results. Systematic verification, validation, and calibration are needed.
Recommendations for performance enhancement	Temptation to provide recommendations without generating adequate confidence in key stakeholders about the model and the modelers (may cause delay or rejection of recommendations)	Involve all the stakeholders in verification and validation exercise; do not project the "model" in "black box" fashion; raise confidence first in modeler and then in a computational model.

References

Al-Abbas, A.H., Naser, J., and Dodds, D. (2012). CFD modelling of air-fired and oxy-fuel combustion in a large-scale furnace at Loy Yang: A brown coal power station, *Fuel*, 102, 646–665.

Arenillas, A., Backreedy, R.I., Jones, J.M., Pis, J.J., Pourkashanian, M., Rubiera, F., and Williams, A. (2002). Modelling of NO formation in the combustion of coal blends, *Fuel*, 81, 627–636.

Asotani, T., Yamashita, T., Tominaga, H., Uesugi, Y., Itaya, Y., and Mori, S. (2008). Prediction of ignition behavior in a tangentially fired pulverized coal boiler using CFD, *Fuel*, 87, 482–490.

Backreedy, R.I., Habib, R., Jones, J.M., Pourkashanian, M. (1999). An extended coal combustion model, *Fuel*, 78(14), 1745–1754.

Barth, R., Meyer, M., and Spitzner, J. (2012). Typical pitfalls of simulation modeling - Lessons learned from Armed Forces and business, *J. Artificial Societies and Social Simulation*, 15(2), 5, http://jasss.soc.surrey.ac.uk/15/2/5.html.

Baum, M.M. and Street P.J. (1970). Predicting the combustion behavior of coal particles, *Combust. Sci. and Technol.*, 3(5), 231–243.

Belosevic, S., Sijercic, M., Oka, S., and Tucakovic, D. (2008). A numerical study of a utility boiler tangentially-fired furnace under different operating conditions, *Fuel*, 87, 3331–3338.

Bezzo, F., Macchietto, S., and Pantelides, C.C., (2003). General hybrid multizonal CFD approach for bioreactor modeling, *AIChE J.*, 49, 2133.

Bhasker, C. (2002). Numerical simulation of turbulent flow in complex geometries used in power plants, *Adv. Eng. Software*, 33, 71–83.

Coda, B. and Tognotti, L. (2000). The prediction of char combustion kinetics at high temperature, *Exp. Thermal and Fluid Sci.*, 21, 79–86.

Crowe, C.T., Schwarzkopf, J.D., Sommerfeld, M., and Tsuiji, Y. (2012). *Multiphase Flows with Droplets and Particles, 2nd edition,* CRC Press, Taylor & Francis Group LLC, Boca Raton, FL.

Díez, L.I., Cortés, C., and Campo, A. (2005). Modelling of pulverized coal boilers: Review and validation of on-line simulation techniques, *Appl. Thermal Eng.*, 25(10), 1516–1533.

Díez, L.I., Cortés, C., and Pallares, J. (2008). Numerical investigation of NO_x emissions from a tangentially fired utility boiler under conventional and overfire air operation, *Fuel*, 87, 1259–1269.

Fan, J.R., Zha, X.D., and Cen, K.F. (2001). Study on coal combustion characteristics in a W-shaped boiler furnace, *Fuel*, 80, 373–381.

Fan, J., Sun, P., Zha, X., and Cen, K. (1999). Modeling of combustion process in 600 MW utility boiler using comprehensive models and its experimental validation, *Energy Fuels*, 13-5,1051–1057.

Fang, Q., Musa, A.A.B., Wei, Y., Luo, Z., and Zhou, H. (2012). Numerical simulation of multifuel combustion in a 200 MW tangentially fired utility boiler, *Energy & Fuels*, 26(1), 313–323.

Filkoski, R.V. (2010). Pulverised-coal combustion with staged air introduction: CFD analysis with different thermal radiation methods, *The Open Thermodynamics J.*, 4, 2–12.

Freitas, C.J. (2002). The issue of numerical uncertainty, *Appl. Math. Modelling* 26, 237–248.

Frey, P. and George, P.-L. (2008). *Mesh Generation, 2nd edition,* Wiley publication, New York.

Gupta, D.F. (2011). Modeling of Coal Fired Boiler, Ph.D. thesis, University of Pune, India.

Hurt, R., Sun, J., and Lunden, M. (1998). A kinetic model of carbon burnout in pulverized coal combustion, *Combustion and Flame*, 113, 181–197.

Liseikin, V.D. (2010). *Grid Generation Methods, 2nd edition,* Springer, Berlin.

Magnussen, B.F. and Hjertager B.II. (1976). On mathematical models of turbulent combustion with special emphasis on soot formation and combustion, in *16th Symp. (Int.) on Combustion*, The Combustion Institute.

Oberkampfa, W.L. and Trucano, T.G. (2002). Verification and validation in computational fluid dynamics, *Progr. Aerospace Sciences*, 38, 209–272.

Ranade, V.V. (2002). *Computational Flow Modeling for Chemical Reactor Engineering*, Academic Press, London.

Roache, P.J. (1997). Quantification of uncertainty in computational fluid dynamics, *Annu. Rev. Fluid. Mech.*, 29, 123–60.

Smith, I.W. (1971). Kinetics of combustion of size graded pulverized fuels in the temperature range 1200–2270°K, *Combustion and Flame*, 17(3), 1971.

Suuberg, E.M. (1991). *Fundamental Issues in Control of Carbon Gasification Reactivity,* Kluwer Academic, Dordrecht, The Netherlands, p. 269.

Thompson, J.F., Warsi, Z.U.A., and Mastin, C.W. (1997). *Numerical Grid Generation: Foundations and Applications, 2nd edition,* North-Holland, Amsterdam.

Tian, Z.F., Witt, P.J., Schwarz, M.P., and Yang, W. (2009). Numerical modelling of brown coal combustion in a tangentially-fired furnace, in *Seventh Int. Conf. on CFD in Minerals and Process Industries,* CSIRO, Melbourne, Australia.

Vuthaluru, R. and Vuthaluru, H.B. (2006). Modelling of a wall fired furnace for different operating conditions using FLUENT, *Fuel Processing Technol.,* 87,633–639.

Williams, A., Backreedy, R.I., Habib, R., Jones, J. M., Pourkashanian, M. (2002). Modelling coal combustion: The current position, *Fuel,* 81(5), 605-618.

Xu, M., Azevedo, J.L.T., and Carvalho, M.G. (2000). Modelling of the combustion process and NO$_x$ emission in a utility boiler, *Fuel,* 79(13), 1611–1619.

Xu, M., Azevedo, J.L.T., and Carvalho, M.G. (2001). Modeling of a front wall fired utility boiler for different operating conditions, *Computer Meth. Appl. Mechanics Eng.,* 190, 3581-3590.

Yin, C., Caillat, S., Harion, J., Baudoin, B., and Perez, E. (2002). Investigation of the flow, combustion, heat-transfer and emissions from a 609 MW utility tangentially fired pulverized-coal, *Fuel,* 81(8), 997–1006.

7

Summary and the Path Forward

Pulverized coal (PC) fired boilers will continue to play an important role in electricity generation for the foreseeable future. Considering the central role of the combustion furnace in such boilers, there is tremendous potential for applying computational models to enhance the performance of such PC fired boilers. Successful modeling of PC fired boiler requires expertise from different fields, ranging from coal chemistry to fluid mixing and transport processes. Computational flow modeling (or CFD) is being increasingly used for improving combustion processes and combustion equipment. We hope that this book conveys the potential of computational modeling for applications to pulverized coal fired boilers and facilitates further improvements in the design, operation, and optimization of such systems.

We have made an attempt to provide adequate information to understand and to define specific aspects of computational modeling for PC fired boilers. The discussion will help one select appropriate models, and apply these computational models to link PC fired furnace hardware to its performance. Because CFD simulations of large boilers are computationally expensive and time consuming, another approach is also proposed based on the reactor network model, which is useful in building lower-order computational models based on data extracted from detailed CFD models. This book described the methodology and process of developing such models for PC fired boilers. More specifically, we have tried to provide

- A methodology for estimating the kinetics of pulverized coal combustion from drop-tube furnace studies. This approach is quite general in nature and can be extended to gasification studies as well.
- Methodology for CFD modeling of PC fired boilers.
- Methodology for developing lower-order computational models from detailed CFD models.
- Guidelines for bridging the gap between available state-of-the-art computational models and their applications in practice.

The book provides guidelines for judicious and effective use of mathematical models in practice, along with some common pitfalls. An attempt was made to evolve general guidelines that may be useful for effective applications of computational modeling to enhance PC fired furnace design and operation. Key points are summarized here. Some of our thoughts on the path forward are also included at the end.

The importance of pulverized coal fired boilers, key issues in understanding and enhancing the performance of these boilers, and the overall approach for harnessing computational modeling for this purpose were discussed in the first 2 chapters of this book. It was emphasized that multiple layers of computational models are more effective for simulating complex systems such as pulverized coal fired boilers. Various steps in the combustion of coal particles were described in Chapter 3. Kinetics of the devolatilization and combustion of coal particles along with the methods to obtain this information were also discussed in that chapter. It is emphasized that kinetics estimated using simple, plug flow-like models may be misleading and may not be appropriate for incorporation into large computational flow models of PC boilers. The chapter also provides adequate suggestions on the use of computational fluid dynamics models to estimate appropriate kinetic parameters from the drop tube furnace data.

Chapter 4 provided a detailed step-by-step approach to develop detailed CFD models for PC fired boilers. Adequate background information and comments for helping with the appropriate selection of various submodels of the CFD model were included. The application of a developed CFD model for simulating 210-MWe tangentially fired boilers was illustrated in this chapter. The developed models and suggested simulation approach will be useful for practicing engineers to develop computational models of PC boilers in an effective way. Simulations based on the CFD model can be used to formulate a multizone or compartment-based phenomenological model that is computationally orders of magnitude less expensive than the CFD model. This was discussed in Chapter 5. Key steps involved in developing such a phenomenological model using the results simulated by the CFD model were discussed and illustrated for the 210-MWe tangentially fired PC boiler. Some tips and comments for facilitating the application of computational models to industrial PC boilers and for effective utilization of such models were included in Chapter 6. Some common pitfalls for the application of models to practice were also discussed. The discussion will be useful to practicing engineers and researchers interested in enhancing the performance of PC boilers. This chapter (Chapter 7), in addition to summarizing the key points discussed in this book, shares some of the authors' thoughts on path going forward.

It is beneficial and more efficient to develop mathematical and computational models in several stages, rather than directly working with and developing a single comprehensive model. For example, even if the objective is to simulate turbulent, multiphase, nonisothermal, and reactive flows in a PC fired furnace, it is always useful to undertake a stage-wise development and validation of computational models. Such stages might include the following:

- Carry out cold air flow simulations.
- Examine these results and select an appropriate turbulence model; carry out simulation of turbulent flow

- Evaluate isothermal, turbulent simulations, verify existence of key flow features, try to quantitatively validate wherever possible.
- Include particle phase (gas-solid flows)
- Include nonisothermal effects (without reactions).
- Include reactions.
- Examine results for qualitative features; validate quantitatively wherever possible.
- Use the models to realize better understanding and achieve performance enhancements.

Development efforts and simulated results from each stage enhance our understanding of the underlying processes. For each stage of model development, quantitative evaluation of limiting solutions (maybe with drastic simplifications) is often useful to enhance confidence in the developed computational model. The simulated results also provide information about the relative importance of different processes, which helps make judicious choices between "simple" and "simpler" representations. Such a multistage development process also greatly reduces various numerical problems, as the results from each stage serve as a convenient starting point for the next stage.

Apart from appropriate model formulation, it is also essential to understand the influence of numerical issues (e.g., grid spacing, time step, degree of convergence, etc.) on simulated results before one can use the results obtained from a computational flow model for engineering applications. One must resist the temptation to use physically realistic simulated results without quantitatively assessing grid dependence. In many practical situations, however, it may not be possible to obtain grid-independent solutions for PC fired furnaces (due to the constraints on available time and computational resources). In such cases, models and simulations can still be used for practical applications, provided some of the following precautionary steps are carried out:

- Quantitative evaluation of special cases/limiting solutions
- Qualitative verification of key flow features
- Assessing dependence on grid spacing by extrapolating key results to zero grid spacing (results may not be grid independent even for the finest grid used in these simulations)

The analyst's paradox is: "Everyone believes an experiment except the experimentalist. No one believes an analysis except the analyst." The onus of establishing the credibility of simulated results is therefore on the modeler. Therefore, model validation and calibration should be given adequate attention. New approaches are needed to quantify the uncertainty in the simulated results. Model calibration focuses on identifying the right parameters that can be adjusted to reduce the observed gap between simulated results and reality. The right parameters are usually identified either by carrying out a sensitivity analysis or are based on experience/engineering judgment. Model calibration is an essential step in applying computational models to practice. It deserves more attention, and systematic methodologies should be developed for calibration exercise.

With the emergence of inexpensive, high-speed computing platforms and the availability of commercial CFD codes and support, flow modeling must be harnessed to devise the best-possible reactor hardware. Some comments on future trends and needs may be appropriate at this juncture. Each advance in the CFD community's capability to perform a particular class of computations has led to a corresponding increase in the engineer's expectations. These expectations can be translated to define research and developmental requirements. These requirements can be classified into two categories: computational and physical.

The most important areas of computational character for which further work is needed include

- Developing ways for carrying out fine-grid computations with complex physics: The existence of nonlinearities and strong coupling among model equations often demands significant efforts for parallelizing computer codes. Newer ways of combining GPU- (graphics processing unit-) and CPU- (central processing unit-) based computations may offer unprecedented advances in our ability to perform multiphase simulations.
- Developing accurate numerical methods without jeopardizing the robustness
- Preserving the order and flexibility in CFD codes as the complexity of their physical content increases

Ranade (2002) has pointed out several suggestions for further work on the development of better physical models to be incorporated into computational models. In particular for the modeling of PC fired boilers, further work on the following aspects is needed:

- Development of kinetic models to capture unburned carbon in ash
- Estimation of various parameters of radiation models
- Interaction of hot solid particles and metal tubes
- Influence of solid particles on pressure drop and heat transfer coefficient

In addition to these specific areas, the following points will deserve more attention in the coming years:

- Develop multiple models addressing objectives on different scales and develop a framework for exchanging information among these models.
- Establish new ways of exchanging information among these models of different scales.
- Develop appropriate verification, validation, and model calibration practices.
- Incorporate CFD, multizone models, lower-order models, parameter estimation, and experimental data in software tools that may be deployed at the plant level for performance enhancement.

It is worthwhile to include brief comments about the available CFD tools and platforms. Usually, monolithic codes, in their attempts to cater to a wider market (spanning different industries and covering widely varying flows), often become unmanageable and difficult to support. The licensing policies and costs of such complex commercial CFD software are often restrictive. It may also be difficult to integrate such general-purpose commercial CFD software and other in-house models/technologies related to specific applications. There is an increasing trend toward using open-source CFD platforms because their design is open for customization. Open-source tools, however, have a steeper learning curve, limited documentation and support, and require significant efforts in establishing acceptance of tools/results. Further efforts in developing best practices based on open-source codes, adequate documentation, and possible GUIs (graphical user interfaces) will lead to enhanced applications and facilitate acceptance of open-source code for industrial applications. We anticipate an increase in the trend toward developing vertical applications based either on open-source or commercial platforms. In such vertical applications, the in-house expertise on specific applications, accumulated experience, and experimental/plant data and best practices for applying computational models can be integrated seamlessly for developing easy-to-deploy computational tools. We hope that the discussion in this book will be helpful in harnessing some of these developments in the future.

Adequate attention to key issues discussed in this book and creative use of computational modeling will make significant contributions to enhance understanding as well as performance of pulverized coal fired furnaces in boilers. The field of computational modeling will continue to evolve and develop further. New advances can be assimilated using the framework discussed in this book. We hope that this book will stimulate applications of computational modeling to pulverized coal fired boilers and other related areas.

Notations

Character / Symbol	Definition	Unit
a	Absorption coefficient of gas	1/m
a_p	Equivalent absorption coefficient due to the presence of particulates	1/m
A	Cross-sectional area	m^2
A_c	Preexponential factor for char oxidation	kg/m^2 s Pa
A_{pn}	Projected surface area of particle "n"	m^2
A_r	Preexponential factor for gas phase reaction "r"	1/s
A_v	Preexponential factor for devolatilization	1/s
C_0	Viscous resistance coefficients	$1/m^2$
C_2	Inertial resistance coefficients	1/m
C_1, C_2	Parameters of k–ε model	—
C_D	Drag coefficient	—
$C_{l,r}$	Molar concentration of each reactant lth species in reaction r	kmol/m^3
C_p	Heat capacity	J/kg K
C_μ	Parameter of turbulence model for estimation of turbulent viscosity	
d	Characteristic dimension of heat exchanger tube	m
d_d	Outer diameter of the deposit	m
d_p	Particle diameter	m
D_{km}	Diffusion coefficient for species k in the mixture	m^2/s
d_l	Outer diameter of the tube	m
E_b	Emissive power of black body	W/m^2
E_c	Activation energy for char oxidation	J/kmol
E_p	Equivalent emission	W m^{-3}
E_r	Activation energy for gas phase reaction "r"	J/kmol
E_v	Activation energy for devolatilization	J/kmol
f_d	Diffuse fraction	
$f(E)$	Distribution curve of the activation energy E	
F	Force acting on the particle	N
F	Mass flow rate	kg/s
f_{heat}	Fraction of char oxidation heat absorbed by the particle	—
f_{pn}	Particle scattering factor associated with the nth particle	—
g	Gravitational constant	m^2/s
G	Turbulence generation term	
G	Incident radiation = $4\sigma T^4$ for the P-1 radiation model	W/m^2
h	Gas enthalpy	J/kg
h	Heat transfer coefficient	W/m^2K

h_k^0	Enthalpy of formation of species k at the reference temperature, T_{ref}	J/kg
h_j^0	Standard heat of formation of species j	J/kg
H	Total enthalpy	J/kg
H_c	Heat released during char oxidation	J/kg
H_r°	Standard heat of reaction	KJ/mol
I	Radiant intensity	W/m²
j	Index of reactions	
k	Thermal conductivity of gas	W/m K
k	Turbulent kinetic energy	
K_c	Char oxidation kinetic rate constant	kg/m² s Pa
K_d	Gas phase diffusion coefficient for oxygen	kg/m² s Pa
K_d	Conductivity of the deposit	W/mK
K_r	Kinetic rate constant for reaction r	1/s
$K_{r,1;}$ $K_{r,2}$ and $K_{r,3}$	Kinetic rate constant for reverse reaction (NO_x)	m³/gmol . s
k_t	Turbulent thermal conductivity of gas	W/m K
K_v	Devolatilization kinetic rate constant	1/s
Kv_i	Devolatilization rate constant	1/s
L	Characteristic dimension of the furnace	m
m	Index for reactions	—
$\dot{m}$	Mass flow rate of particle	kg/s
M_f	Final mass of sample after the devolatilization is complete	kg
m_k	Mass fraction of species k	—
M_k	Molecular weight of species	k
m_j	Mass fraction of species j	_
M_p	Mass of particle	kg
M_v	Mass of volatile at any time	kg
m_v, m_c	Mass fraction of volatile and char respectively	—
$m_{v,0;}$ $M_{w,0}$	Mass fraction of volatiles and moisture initially present in the particle	
$MW, M_{w,k}$	Molecular weight	kg/kmol
n	Order of reaction	—
N_p	Number of particles	1/s
N_r	Number of reactions	—
Ny	Number of experiments	—
Nz	Total number of data points for each experiment	—
p	Static pressure	Pa
$P_{O_2}^n$	O_2 partial pressure	Pa
Pr	Prandtl number	—
q_{in}	Total radiant intensity on the walls	W/m²
Q_{char}	Heat of reaction of char oxidation	W
Q_r	Radiative source term	W/m³
Q_{cond}	Heat conducted through the porous deposit	W
Q_{conv}	Heat convection to the deposit from flue gas or heat flow towards the particle by convection	W

Q_{emi}	Heat emitted by the deposit	W
Q_{rad}	Heat radiated from the gas and other emitting bodies or heat flow toward the particle by radiation	W
r	Radial direction, any position in the computation domain	
Re	Reynolds number	
R_k	Net rate of production of species k by chemical reaction	$kg/m^3\,s$
R_j	Net rate of production or consumption of species j by chemical reaction	kg/s
R_{max}	Maximum vitrinite reflectance	
R_v	Rate of devolatilization	kg/s
R_c	Rate of char oxidation	kg/s
s	Path length in radiative heat transfer equation	m
Sc_t	Turbulent Schmidt number, $Sc_t = \frac{\mu_t}{\rho D_t}$	
$Sgas\text{-}rxn$	Source term of heat of chemical reactions	W
$Srad$	Heat transfer by radiation from all other zones	W
$Schar$	Source term for discrete phase char oxidation	W
$S_{h,rxn}$	Source term for heat of gas phase chemical reactions	W/m^3
S_k	Source of species k from dispersed phase	$kg/m^3 s$
S_j	Source of species j from dispersed phase	kg/s
S_m	Source term for the total mass added from the discrete phase	$kg/m^3 s$
S_Q	Source term for heat added from discrete phase	W/m^3
St	Stokes number	
St_{eff}	Effective Stokes number (for particle deposition model)	
S_U	Source term for radial momentum	$kg/m^2\,s^2$
S_W	Source term for axial momentum	$kg/m^2\,s^2$
T_d	Temperature of the deposit	K
T_g	Local gas temperature	K
T_m	Melting temperature of particle	K
T_p	Particle temperature	K
T_R	Radiation temperature $= \left(\frac{I}{4\sigma}\right)^{1/4}$	K
$T_{ref,k}$	Reference temperature	K
T_t	Tube metal temperature	K
U	Fluid velocity in radial direction r	m/s
U	Instantaneous fluid velocity	m/s
$\overline{U}$	Time average velocity	m/s
U	Unburned fraction of coal	
u	Fluctuating velocity	m/s
U_p	Velocity vector of the particle	m/s
$u_{p,i,j}$	Velocity components of the particle of size j in i^{th} direction (r or z)	m/s
v_i	Gas velocity in i^{th} direction (U or W)	m/s
V	Cell volume	m^3
V	Volume of nth internal CSTR of kth zone	m^3
Vp	Particle volume (Equation (5.12))	Vp

V	Volatile yield as function of temperature	%
V^*	Total amount of volatile yields	%
W	Fluid velocity in axial direction z	m/s
w	Weighting factor	
wj	Initial mass of particle of size j	kg
y_{exp}	Experimental value	
Y_p	Mass fraction of any product species, p	
Y_R	Mass fraction of a particular reactant, R	
y_{sim}	Calculated value	
z	Axial direction	
z	Number of data point for each experiment	

Subscript

0	Initial condition	
A	Ash present in coal	
b	Blackbody	
c	Char	
c	Char present in coal	
d	Diffuse (for radiative heat transfer equations)	
d	Drag	
db	Dry basis	
g	Gas	
i	Direction or reaction or component, trajectory of particle	
j	Particle size or number of pseudo components	
k	Species	
O_2	Oxygen	
p	Particle	
Q	Discrete phase	
r, R	Radiation	
s	Direction	
t	Turbulent	
v	Volatile present in coal	
w	Moisture present in coal	
w	Wall (for radiative heat transfer equations)	
λ	Wavelength	

Greek letter

μ	Viscosity of gas phase	kg/m s
μ_i	Turbulent or eddy viscosity	kg/m s
μ_{eff}	Effective viscosity	kg/m s
δij	Kronecker delta function	
σ_p	Equivalent particle scattering factor	
σ	Stefan–Boltzmann constant = 5.67×10^{-8}	$W/m^2 K^4$
	Standard deviation in Figure 3.7	kJ/mol
ε_{pn}	Emissivity of particle	

ε_w	Emissivity of wall	
ε	Turbulent energy dissipation rate	m^2/s^3
Δ	Change in the property across that cell	
ρ	Density	kg/m^3
$v'_{k,r}$ and $v_{k,r}$	Stoichiometric coefficient for k^{th} species as product and reactant respectively	
$\Phi, \phi(\vec{s}, \vec{s}')$	Scattering phase function	
$\eta'_{l,r}$	Exponent for each lth reactant in reaction r	
β	Heating rate	K/min
Ω	Solid angle	Radian
ω	Direction of propagation of intensity	
φ	Factor for estimating effective Stokes number	
ε_d	Emissivity of the deposit	
ϵ	Emissivity of the gas	
$\phi, \alpha, \beta,$ and γ	Model parameters for estimation of impact efficiency	
η_{ref}	Critical viscosity of slag	Pa-s
η_p	Particle viscosity	Pa-s
η_s	Sticking efficiency	
η_i	Impact efficiency	
τ_{fL} and l_e	Characteristics time and length scales of turbulence eddies	s and m
$\Phi_{k,k}$	Parameter that connects the outlet of all zones to current k^{th} zone	
θ	The time spent by the particle in each zone	s
$\tau_{k,n}$	Residence time of the particle in any n^{th} internal reactor of kth zone	s

Abbreviations

CBK	Char Burnout Kinetic model
CFD	Computational fluid dynamics
daf	Dry ash-free basis
DPM	Discrete particle model
EE	Eulerian–Eulerian
EL	Eulerian–Lagrangian
FC	Fixed carbon/char
FEGT	Furnace exit gas temperature
FT	Flow temperature
GCV	Gross calorific value
HT	Hemispherical temperature
IDT	Initial deformation temperature
IGCC	Integrated gasification combined cycle
LECO	Lower economizer heat exchanger
LES	Large eddy simulations
LTSH	Low temperature superheater
MWe	Mega Watt electric

MW_{th}	Mega watt thermal
Pass2TOP	Volume above LTSH
PC	Pulverized coal
PDF	Probability density function
PSD	Particle size distribution
RANS	Reynolds averaged Navier–Stokes equations
RMS	Root mean square error
RNG	Renarmalized group
RNM	Reactor network model
RSM	Reynolds stress model
RTD	Residence time distribution
SC	Supercritical
ST	Softening temperature
UBC	Unburnt char (or carbon)
UECO	Upper economizer heat exchanger
USC	Ultra-supercritical
V_m	Volatile matter
WSGGM	Weighted sum of gray gas model

Index

A

Acid rain, 110
Air inlets, 67, 170, 211–214
Ambient conditions, 3
American Society for Testing of
  Materials, 33, 45
Amine compounds, 111
Ammonia, 111
Application of computational models,
  29–30
Ash composition, Indian coal, 35
Ash content, 16, 34, 36, 40, 43, 113, 153
Ash fusion temperature data, 36
ASTM. *See* American Society for
  Testing of Materials
Australia, coal-based electricity, 1

B

Bed combustion, 6–7
Boiler configuration, 19, 24, 26, 28
Boiler downtime, 208
Boiler-level models, 5, 15, 23, 26–30, 57,
  75, 81, 155, 157, 165, 238, 247
Boiler performance enhancement, 11
Boundary conditions, computational
  fluid dynamics models,
  212–215
Burner inlets, 56, 182, 186, 212–214
Burner tilt, 16, 119, 130, 143, 145–150, 157,
  167–168, 173–175, 177–179
Burner zone, 9, 25, 135, 143, 169
  it can also be called combustion
    zone, 169

C

CAVTs. *See* Cold air velocity tests
CBK. *See* Char burnout kinetic
CFD. *See* Computational fluid
  dynamics

CFM. *See* Computational flow
  modeling
Char burnout kinetic model, 97, 247
Char burnout models, 44
Characteristics of fuel, 19
Chemical percolation devolatilization
  model, 38–39, 93
Chemical reaction engineering, 13, 30,
  165
China, coal-based electricity, 1
CO/d2/D emission, 3
Coal-based electricity, 1
Coal blends, 42, 75, 119, 143, 152–153,
  155–158, 234
Coal characterization, 19, 34, 45–53
Coal devolatilization/combustion
  kinetics, 33–80
  ash composition, Indian coal,
    35
  ash fusion temperature data,
    36
  char burnout models, 44
  coal characteristics, 34
  coal characterization, 45–53
  data processing, 47–53
  drop-tube furnace, 53–74
    boundary conditions, 67–68
    data processing, 55–74
    numerical simulations, 67–68
    parameters, 67
  experimental techniques, 41–43
  non-coking coal, 36
  thermogravimetric analysis, 45–53
Coal fired boilers, 6–12, 19–32
  application of computational
    models, 29–30
  bed combustion, 6
  boiler-level models, 26–29
    computational fluid dynamics
      models, 5, 15, 23, 27–28, 30, 57,
      75, 81, 155, 247
    engineering design models, 23–24

Printed and bound by CPI Group (UK) Ltd, Croydon, CR0 4YY

22/10/2024

01777647-0002